Robert Oliphant Pringle

The Diseases of Horses, Cattle, Sheep, Swine, Dogs, and Poultry,

Their Causes, Symptoms, and Treatment

Collected and arranged from the best authorities

Robert Oliphant Pringle

The Diseases of Horses, Cattle, Sheep, Swine, Dogs, and Poultry, Their Causes, Symptoms, and Treatment
Collected and arranged from the best authorities

ISBN/EAN: 9783337328832

Printed in Europe, USA, Canada, Australia, Japan

Cover: Foto ©berggeist007 / pixelio.de

More available books at **www.hansebooks.com**

THE DISEASES

OF

HORSES, CATTLE, SHEEP, SWINE, DOGS,

AND

POULTRY,

THEIR CAUSES, SYMPTOMS, AND TREATMENT:

COLLECTED AND ARRANGED

FROM THE BEST AUTHORITIES

BY

R. O. PRINGLE,

EDITOR "IRISH FARMERS' GAZETTE," AUTHOR OF "MEMOIR OF WILLIAM
DICK," "THE MEAT MANUFACTURE," "PURDON'S
PRACTICAL FARMER," ETC.

DUBLIN:
PURDON, BROTHERS, 23, BACHELORS'-WALK
WILLIAM BLACKWOOD AND SONS, EDINBURGH AND LONDON.
1871.

INTRODUCTION.

THE object of this work is to afford information relative to the diseases of domesticated animals, to those who may not possess the advantage of being within reach of good professional advice, and also in cases of emergency, even where such advice can be obtained, but where immediate treatment may be absolutely neces- sary to save the life or alleviate the sufferings of any of those animals which are dependant upon our care and attention.

Before proceeding to describe the causes, symptoms, and treat- ment of the various maladies to which the live stock of the farm and other kinds of domesticated animals are liable, we are anxious that our readers should be fully impressed with the fact that most of those diseases are capable of being prevented by the exercise of ordinary care and forethought. This will be clearly seen when the causes of disease, as detailed in the following pages, are taken into consideration ; and we would therefore press the importance of attending closely to many points in management, simple enough in themselves, but the neglect of which not unfrequently leads to very disastrous consequences.

When we turn, for instance, to the maladies which affect the horse, and consider the causes by which they were produced, how many do we find the direct result of neglect, of ignorance, and even of downright cruelty ! Bad food, feeding and watering at improper times or in an improper manner, filth, want of ventila-

tion in stables, blows, and other kinds of harsh treatment; and, on
he other hand, mistaken kindness, fashion, vanity, and a variety
of other causes, all of which are under the control of man, com-
bine to ruin the health, destroy the usefulness, and even terminate
the existence of the noblest animal which has been placed at the
service of man.

It is the same when we take into consideration the diseases to
which other classes of domestic animals are liable. The causes
which produce them are also, for most part, preventible, or capable
of being removed if already in existence ; and we trust, therefore,
that every owner of live stock into whose hands those pages may
come will never forget the important truth conveyed in the old
proverb—" Prevention is better than cure."

Our experience in conducting agricultural journals has long
since satisfied us that a Hand Book of reference, in which the
causes, symptoms, and general treatment of the various diseases
incident to the domesticated animals would be detailed in a clear
but concise manner, is much required for the use of the owners of
such animals, and those entrusted with the care of them. In preparing this work the best writers have been consulted, and the
information so obtained condensed as much as possible, without
affecting the utility of the work. Veterinary information is scat-
tered throughout a number of publications, most of which are ex-
pensive, and few owners of stock, therefore, are possessed of a sufficient number of works of that kind to refer to in all emergencies.
We believe, therefore, that this " Veterinary Hand Book" will
meet the wishes and wants of a large number of persons, includ-
ing, as it does, the causes, symptoms, and treatment of the diseases

of HORSES, CATTLE, SHEEP, SWINE, DOGS, and POULTRY, with information relative to certain accidents to which animals are occasionally liable, and operations which must be performed upon them.

In order to show the authorities which have been consulted in preparing this work, we give the following list :—

YOUATT on the Horse (Last Edition).

       Do.   Cattle.

       Do.   Sheep.

       Do.   Swine (Sidney's Edition).

Lieut.-Col. FITZWYGRAM's "Horses and Stables."

Professor DICK's Veterinary Manual.

       Do.   "Veterinary Papers" (Blackwood and Sons).

Professor HUGH FERGUSON's (H.M.V.S. Ireland) Official Reports and Communications, &c.

FINLAY DUN's Prize Essays, &c.

GAMGEE's "Dairy Stock."

      Do. "Domesticated Animals," &c.

      Do. "Veterinarian's Vade-Mecum."

Professor MORTON's "Veterinary Pharmacy."

Professor TUSON's "Veterinary Pharmacopœia."

W. C. SPOONER in Morton's Cyclopedia.

MAYHEW's "Horse Doctor."

      Do. on Dogs.

"MAGENTA'S" Domesticated Dogs (Blackwood & Sons).

ARMATAGE's Clater's Cattle Doctor.

Prize Essays and Transactions of the Highland and Agricultural Society (Blackwood and Sons).

Journal of the Royal Agricultural Society of England.

Professor SIMONDS' Official Reports and Lectures.

R. H. DYER, V.S., Communications to *Irish Farmers' Gazette,
    Veterinarian,* &c.

Professor LAW's Occasional Papers in the *Scottish Farmer,* &c.

Poultry Chronicle (*Journal of Horticulture*).

The *Field.*

The *North British Agriculturist.*

The *Irish Farmers' Gazette.*

Bell's Messenger.

The *Mark-lane Express.*

The *Veterinarian.*

Agricultural Gazette.

&c.   &c.   &c.

CONTENTS.

CHAP. VI.

THE SKIN.

CHAP. VII.

THE LIMBS AND FEET.

CHAP. VIII.

SPECIAL DISEASES.

PART II.

DISEASES OF CATTLE.

CHAP. I.

THE HEAD.

CHAP. II.

THE THROAT AND RESPIRATORY ORGANS.

CHAP. III.

The Stomach, Liver, Spleen, Bowels, Kidneys, Blood.

CHAP. IV.

Parturition—The Udder, Etc.

CHAP. V.

The Skin.

CHAP. VI.

Muscular System and Extremities.

CHAP. VII.

Special Diseases.

PART III.

DISEASES OF SHEEP.

CHAP. I.

The Head and Nervous System.

PART IV.

DISEASES OF SWINE.

CHAP. I.

THE HEAD, NERVOUS SYSTEM, AND ORGANS OF RESPIRATION.

CHAP. II.

THE INTERNAL ORGANS.

CHAP. III.

THE SKIN.

PART VI.

DISEASES OF POULTRY.

Pages 189—194.

PART VII.

GENERAL SUBJECTS.

CHAP. I.

Wounds, Fractures, Etc.

CHAP. II.

Operations, Applications, &c.

Appendix.

THE
VETERINARY HAND-BOOK.

PART I.

DISEASES OF HORSES.

CHAPTER I.

THE HEAD AND NERVOUS SYSTEM

APOPLEXY.

Causes.—Sudden determination of blood to the head; high feeding, without exercise; pressure of collar.

Symptoms.—The horse is sometimes struck down as if shot, and loses all sense and power of motion. At other times the animal keeps his head low, he staggers as he stands, and appears as if he would fall; sight and hearing are affected. After a time he falls; grinds his teeth; eyes open, protruded, and fixed; there are twitchings about the frame, which increase to convulsions; he is unable to swallow, and soon dies.

B

Treatment.—Copious bleeding, with stimulating mustard embrocations to the belly and spine. There is, however, little hope of effecting a cure.

EPILEPSY.

Cause.—This is an aggravated form of the malady known as megrims, vertigo, or giddiness (*see same*).

Symptoms.—The horse rears up and falls suddenly, or he reels about and falls; the muscles of the eye are affected with spasm, so that this organ is greatly distorted; the breathing is disturbed, and there is sometimes a violent motion of the legs. The fit may last only a few minutes, or it may extend over several hours.

Treatment.—Bleed, if the horse is plethoric; keep the bowels open; feed moderately; insert setons in the neck; but, as Mr. Youatt says, "he who values his own safety, or the lives of his family, will cease to use an epileptic horse."

LOCKED JAW.

Causes.—Irritation of a punctured wound in some tendenous part; exposure to cold and wet, and sometimes apparently spontaneously, especially in hot climates.

Symptoms.—Spasms of the muscles of the jaw and neck, which extend to the back and loins. The animal does not feed; he drops his food and gulps water. The jaws become closed, the head poked out, neck rigid, nostrils dilated, and back quite stiff, so that the animal moves as it were in a piece.

Treatment.—Look for the wound which has preceded the disease, and if there be irritation, relieve it. Bleed copiously, and give powerful purgative medicines, and croton oil will be the most easily administered. Give of this 8 to 12 drops in a pint of castor oil, followed by opium and camphor, in doses of a drachm each. Rub well the jaw and neck, and also the back and loins, with turpentine, soap liniment, mustard, &c. Blistering the belly has also sometimes proved of service. Repeat the oil, without the croton, if necessary. Put a pailful of thin gruel within reach, and keep the animal in a state of perfect quietude, as the least excitement is injurious. Should the animal recover, careful nursing for some time will be necessary.

MEGRIMS, VERTIGO, OR GIDDINESS.

Causes.—Momentary congestion of the brain, tight collars, tight reining up and bearing reins, assisted by hot weather and bright sun.

Symptoms.—The animal suddenly shakes and throws up his head, or shakes it violently; reels; then stands for a minute or two dull and listless, or runs round and falls to the ground, remaining for a few moments partially insensible, or in a state of violent convulsion. The attack rapidly passes away, the horse rises in a minute or two, shakes himself, and proceeds as if nothing had happened. The attacks may become periodical.

Treatment.—Careful attention to diet, easy, regular work, and occasional physic; see that the collar fits easily, and leave off bearing reins; but a horse subject to giddiness is not to be depended on—certainly not in harness.

PARALYSIS.

Causes.—A slip or fall, causing severe injury to the spine; violent exertion.

Symptoms.—The horse cannot stand, or one hind leg gets in the way of the other, and threatens to throw the animal down.

Treatment.—In mild cases apply a fresh sheep skin to the loins, with the woolly side out; follow this by blistering the spine, and keep the bowels open. Give the following ball, night and morning:—Strychnia, $\frac{1}{2}$ grain, which may be gradually increased to a grain and a half; iodide of iron, one grain; quassia powder and treacle, a sufficiency.

Treatment is useless in acute cases.

RABIES.

Cause.—Bite of a mad dog.

Symptoms.—" Mad Staggers" have frequently been mistaken for this disease; but in true rabies the animal is not merely frantic, but positively and wilfully mischievous, and purposely attacks everything dead or living, which is not the case in " Mad Staggers," the chief characteristic of which is simply furious delirium.

Treatment.—Cut out the bitten part, and apply the hot iron to it. Mr. Spooner states that he has always been successful in applying nitrate of silver after cutting out the bitten part.

STAGGERS.

This is sometimes spoken of as being two distinct diseases, namely, " Sleepy Staggers" and " Mad Staggers;" but veterinarians now consider these distinctions to belong to two stages of the disease, which usually commence with the sleepy stage, although that may merge quickly into " Mad Staggers."

Cause.—Derangement of the organs of digestion, brought on by gorging the stomach with indigestible food, especially after a long fast.

Symptoms.—Dulness and listlessness, loss of appetite, and redness of the eyelids. The animal stands drowsily, and sometimes falls asleep when feeding. These symptoms may continue for some days, and may at last end fatally, or they may be succeeded by wild and furious delirium. The pulse, which in the sleepy state was slower than natural, becomes quicker and smaller, and respiration also quickens.

Treatment.—In the early stage, a large dose of active purgative medicine, as everything depends on getting the bowels to act. Give a pint or a pint and a half of castor oil, with 4 to 6 drops of croton oil. If this does not act in six hours, give a similar dose, without the croton oil, and in the meantime throw up injections of warm water, to which a solution of aloes may be added. One to two drachms of carbonate of ammonia may also be given to assist the action of the purgative. Foment and hand-rub the belly, and rub the legs all over with turpentine liniment. If the disease appears to approach the furious stage, bleed to exhaustion; but after it has fairly passed into the mad stage, no treatment is available, as it is impossible to approach the animal.

STRINGHALT.

Cause.—Obscure; some affection of the nerves.

Symptoms.—Raising both hind legs, one after the other, previous to walking. The disease generally increases with age.

Treatment.—Incurable.

CHAPTER II.

THE EYE.

✗ GUTTA SERENA,

DESIGNATED ALSO AMAUROSIS AND GLASS-EYE.

Causes.—Paralysis of a part or the whole of the optic nerve, produced by excessive glare or heat, or from a pressure on it, such as that produced by the formation of a tumour, by extravasation of blood, or any morbid effusion. A fall backwards or a blow on the head will bring on the disease. It sometimes arises from an over-loaded stomach affecting the nervous system generally, and it also occasionally follows specific ophthalmia.

Symptoms.—Fixed dilitation of the pupil, which is more than ordinarily bright; a ghastly stare in the eye; and the usual symptom of total blindness, such as active ears, restless nostrils, and high, careful stepping.

Treatment.—If the disease proceeds from an overloaded stomach, give a dose of physic, such as powdered aloes, 3 drachms; powdered gentian, 3 drachms; treacle, sufficient to make a ball. In all cases apply cold applications to the head, and put the patient in a dark, well-ventilated box; give cooling, laxative food; but for the most part the case is hopeless.

INFLAMMATION OF THE EYE (OPHTHALMIA).

This may be simple or chronic; and we shall take each kind in succession.

SIMPLE OPHTHALMIA is an inflammation of the membrane which lines the lids and front of the ball of the eye.

Causes.—Introduction of a foreign body, such as particles of sand, or a hair; also exposure to the weather; strong ammoniacal vapours engendered in filthy stables; accompaniment of catarrh and influenza.

Symptoms.—A closing of the eye, and a profusion of tears. On examination the animal closes his eyelids firmly—he cannot bear the light; redness of the membrane; the front part of the ball sometimes appears blue, but in other cases it remains clear and bright.

Treatment.—Examine for any foreign body, which will be generally found under the upper eyelid; and if such be present, remove it. If produced by violent blows over the eyes, there is no chance of a cure. In other cases place the animal in a darkened stall, and tie his head up to the rack. In the early stage apply fomentations of warm water; but when the inflammation has subsided, cold water dressings should be substituted. If the eye remains weak, apply the following wash, once or twice a day :— Sugar of lead, 1 drachm; tincture of opium, 1 drachm; water, 2 pounds; or, Goulard's extract, 2 drachms; tincture of opium,

2 drachms; pure or distilled water, 1 pint. Mix. A dose of physic may be given, and cooling food, substituting bran mashes for corn. Should the disease not yield to simple treatment of this kind, it may be taken for granted that the malady is of a more serious nature, and an experienced veterinary surgeon should at once be consulted.

SPECIFIC OR CHRONIC OPHTHALMIA is variously termed periodical, constitutional, or hereditary ophthalmia; and in stable language, " moon-blindness."

Causes.—Foul air; neglect; cold stables, and general bad stable management. It is also hereditary, and animals affected with it should, on no account, be bred from.

Symptoms.—In early stages same as in common ophthalmia. Then the lids become much swollen; the flow of hot tears is increased; the eye becomes sunk in its socket; there is great redness of the membrane, and blood vessels appear in it; some running in a circular direction, and others radiating from a central point; there is a deposit of watery humour, which becomes thick and muddy, and often tinged with blood: this fills up the eye, so that the interior can no longer be seen. The attack will go off, but it returns again and again after short intervals, until the eye becomes quite disorganised, and cataract is formed; after which the inflammation generally leaves the eye, and does not return. During the attacks, the presence of fever is indicated by the pulse, by dryness of the mouth, constipation of the bowels, and scantiness of urine; but the appetite is seldom affected.

Treatment.—No very great results can be expected from local treatment. Bleeding from the neck vein will be useful when there is much fever; but local bleeding, setons under the eyes, and blisters to the face and jaws, as frequently recommended, have seldom any beneficial result. Large doses of calomel and opium—60 grains of the former and half a drachm of the latter—twice a day, for several days in succession, as soon as effusion has taken place, are recommended. Put the patient into a cool, well-ventilated, but darkened loose box; keep the body warm by clothing, and place a linen shade, kept constantly wet, over the eyes. Good stable management, carefully regulated exercise, cooling diet, and the administration of tonic medicines must be attended to—such as quinine, $\frac{1}{2}$ to 1 drachm, dissolved in a few drops of sulphuric acid, and a pint of water.

CATARACT.

Causes.—The result of specific ophthalmia; injuries; at other times obscure.

Symptoms.—In partial or false cataracts, specks are seen in the centre of the eye, almost the size of a pin's head; but not so dense in appearance as in true cataract. These specks often disappear suddenly. In true cataract, horses see imperfectly, and are apt to shy; for which reason, when an animal evinces a habit of that kind his eyes should be carefully inspected. Cataract, when complete, is shown by the pupil of the eye becoming white, and by total blindness.

Treatment.—Incurable; but a blind horse is generally more useful and safer than one that is partially blind.

TUMOUR IN THE EYE.

Cause.—Unknown.

Symptoms.—A yellow metallic or sea-green appearance in the eye; blindness.

Treatment.—No treatment of any service.

CHAPTER III.

THE MOUTH AND NOSTRILS.

INJURIES TO THE MOUTH.

Causes.—Abuse of the bit ; too tight a curb chain.

Treatment.—In simple injuries of the mouth, a little alum water will be a desirable application, or the wound may be dressed with a little tincture of myrrh, on a piece of tow. The following is also a useful lotion :—Powdered alum, 2 drachms; sulphate of zinc, $\frac{1}{2}$ drachm ; treacle or honey, 1 ounce ; warm water, 12 ounces ; mix. If the bone is injured, so that there is an unpleasant smell from it, inject the chloride of zinc lotion, and keep the wound open, so that the injured bone may come away.

CANKER IN THE MOUTH

Is a disease affecting the gums, particularly the parts under the tongue, and causing a very offensive smell.

Apply the astringent lotion as above.

INJURY TO THE TONGUE.

Causes.—Brutality in giving a ball, whereby the tongue is pulled forcibly out ; also from tying the halter round the tongue and lower jaw, and then leaving the horse tied up, or leading him in hand.

Treatment.—It may be necessary to cut off a part of the tongue ; feed on gruel and soft food, and after each meal wash out the mouth with the chloride of zinc lotion, or the following wash :—Borax, 5 ounces ; honey or treacle, 2 pints ; water, 1 gallon ; mix.

APHTHA, OR APTHOUS THRUSH.

Cause.—Unknown.

Symptoms.—Small swellings on the inside of the mouth, tongue, and lips.

Treatment.—Mashes; a dose of one to two pints of linseed oil; dress the mouth, &c., with the wash recommended for injury to the tongue, as above.

LAMPAS.

Mayhew briefly describes this as " a groom's fancy," which is correct. Professor Dick says it " is often described as a painful swelling of the lower bars of the palate, projecting above the surface of the front teeth, and interfering with the feeding, being a disease of young horses, connected with the shedding of their teeth, and occasioning fever. It is not, however, so much a disease as a natural and salutary process, which, in general, is best let alone, and in which cruel remedies—such as firing—should never for a moment be thought of."

Treatment.—Let the horse have soft, cooling diet, which is all that is necessary.

NASAL GLEET.

Causes.—Neglected catarrh or influenza; also inflammation and thickening of the lining membrane of the nose; diseased teeth; kicks or blows on the face or the nose; existence of polypus or a growth in the nose.

Symptoms.—A chronic discharge, generally intermittent, but sometimes continuous, of matter from one or both nostrils. The discharge falls freely away from the nostrils, and is not of the glue-like character which is peculiar to glanders. It is usually white, and about the thickness of cream, sometimes clotty or lumpy, but generally uniform. Occasionally it is yellowish in tinge, and when caused by disease of the teeth, emits a fetid smell. Although apparently the discharge is suspicious, it is clearly distinguishable from glanders.

Treatment.—Steam the head, and blister over the part affected. Throw cold water up in the nostrils twice a day by means of a large syringe. A solution of sulphate of zinc, in the proportion of one drachm to a pint of water, may also be injected into the nostrils twice a day.

It may be necessary to resort to a surgical operation for the purpose of opening the sinus; but this can only be done by a competent professional man.

When nasal gleet becomes fairly established, little reliance can be placed on treatment. Improve the system by giving mineral and vegetable tonics, such as sulphate of iron or copper in half-drachm doses, with two drachms of extract of gentian, to be given daily for ten days, and then after the interval of a week repeated. Attend to the general health of the animal, and work moderately, but not violently; but it is questionable whether it is safe to keep a horse with an unhealthy discharge from his nostrils, as it is impossible to say when such a case may run into glanders; and as the animal cannot safely work with or be kept among others, it is better to destroy him, which may be done instantaneously, without pain to the animal, by giving him an ounce of prussic acid.

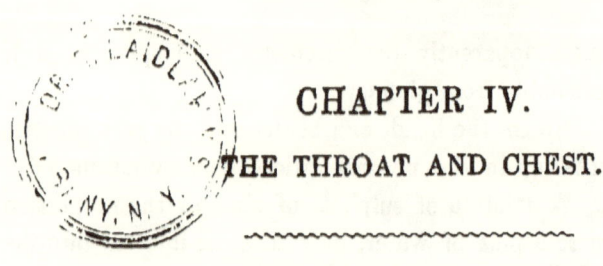

CHAPTER IV.

THE THROAT AND CHEST.

CHOKING—OBSTRUCTIONS IN THE GULLET.

Causes.—A physic ball, or a piece of carrot or turnip, sticks in the gullet.

Treatment.—Use a well oiled, flexible probang; but if this does not succeed the gullet must be opened, but that is an operation which only a skilful veterinary surgeon should attempt. It would be well, in the first instance, if there is not much danger of suffocation, to pour down a little oil, which will tend to lubricate the gullet and induce the obstruction to slip down; when this is done, draw the hand gently and steadily over the outside of the throat. Feed cautiously for some time after on soft food.

CATARRH, COUGH, OR COMMON COLD.

Cause.—Neglect in the management of the animal or the ventilation of stables; young horses when first brought into stables are subject to it.

Symptoms.—Loss of appetite; dulness of the eye; staring coat; a tendency to sweat upon slight exertion; and a little watery discharge from the nostrils. These symptoms are followed by slight feverishness; slightly quickened pulse; somewhat hurried breathing; a hot mouth; bowels usually constipated; increased discharge from the nostrils.

Treatment.—A cool horse box, with abundance of fresh air, extra warm clothing, flannel bandages to the legs; no corn; warm mashes; laxative diet. If the running at the nose is considerable and the cough troublesome, steam the head frequently during the day; if the patient is feverish, a dose consisting of half an ounce of sweet spirits of nitre and two drachms of nitrate of potass may be given once or twice a day for two or three days: if the bowels are constipated, instead of the previous medicine, give a dose of two ounces of Epsom salts, with half an ounce of nitrate of potass, twice a day, until the desired effect is produced. If there is depression, a staring coat, and unequal heat in the legs, one ounce of spirits of nitric ether and four ounces of acetate of ammonia may also be administered in a pint of water, morning and evening.

Cough Ball for Horses.—Take of extract of belladonna half a drachm to 1 drachm; powdered aloes, 1 drachm; nitrate of potass, 2 drachms; common mass to form a ball, which may be given every other day until four balls have been administered.

CHRONIC COUGH.

Causes.—Morbid sensibility of the nerves of the windpipe, or from irritability of the bronchial tubes, after pneumonia, bronchitis, or influenza. It may also arise from indigestion, and horses suffering from worms are often affected with chronic cough. It accompanies broken wind, and sometimes it exists without any apparent cause.

Symptoms.—Chronic cough, when following bronchitis or influenza, is usually accompanied by an extra secretion of mucus; but it also exists when the membrane is particularly dry.

Treatment.—If it arises from irritability of the windpipe, blister the throat, or insert a seton, which is better, as the action may be maintained for any length of time, while the horse may be worked as usual. Mix boiled linseed with the food; give five or six pounds of carrots daily, or steamed food if the horse is employed in slow work. Give the food and water often, but in small quantities at a time.

If the cough proceeds from indigestion, regulate the diet carefully, give abundance of pure air, and occasional alterative medicines, such as nitre, black antimony, and sulphur, of each 2 drachms.

If the cough follows bronchitis, pneumonia, or influenza, and is accompanied with an extra secretion of mucus, with occasional discharge from the nose after coughing, or with a wheezing noise, give a drachm of sulphate of copper with 2 drachms of extract of gentian daily for a week, but if the cough continues, fumigate the box by putting some tar in an iron ladle, and plunging a bar of hot iron into it. Fumigation with sulphur—"the sulphur cure" —is also useful.

Much benefit in most cases of chronic cough will be derived from the use of tar either in the water or in balls. For the former, pour a quart of the best Archangel tar into a large cask, from which the water used for drinking may be drawn as required ; or 2 drachms of tar may be made up into a ball with gentian, and given daily. Horses affected with chronic cough may continue useful for years, if properly fed and carefully attended to in regard to stable management.

BROKEN WIND.

Causes.—The result of pneumonia and other diseases of the respiratory organs ; chronic cough and chronic indigestion, accompanied by excessive distension of the stomach, arising from the use of innutritious or bad forage, such as mildewed or mow-burnt hay, damp or sprouting oats, or stale green food ; or from working a horse on a full stomach, or with his belly full of water.

Symptoms.—Short, dry, hacking cough ; respiration performed by a triple effort ; inspiration spasmodic and single ; expiration laboured and double ; ravenous appetite ; great thirst ; frequent disposition to expel wind by the fundament ; dung half digested ; belly hangs down ; coat ragged ; aspect dejected.

Treatment.—Incurable ; but capable of being much relieved by care in feeding, watering, and working. Never leave idle for a day, and let the food be highly nutritious and concentrated, such as oats and beans ; give little hay ; no straw, and water in small quantities. ᵢ Carrots form good food for a broken-winded horse ;

never work on a full stomach. ✗ In each draught of water, say each half-pail, which is enough at a time, put half a drachm of sulphuric acid.

BRONCHITIS.

Causes.—Same as in " Catarrh" and " Sore Throat," which see.

Symptoms.—Quickened breathing, accompanied with a slight whistling or hissing sound heard on placing the ear to the sides of the chest, or else by a deeper and more noisy sound in front of the chest. Increase of the attack is marked by hurried breathing, dilatation of the nostrils, heaving of the flanks, much fever, and prostration of the strength.

Treatment.—See generally treatment in " Sore Throat." At first it may be advisable to give an ounce of spirits of nitric ether, with four ounces of acetate of ammonia in eight ounces of water, both morning and evening, and a ball consisting of two drachms each of resin, nitre, and antimony. If this does not avert the attack, give five to ten drops of Fleming's tincture of aconite every four hours, but this must be discontinued after the pulse has become soft, or say in about 24 hours. When the patient becomes prostrated, give diffusible stimulants, such as carbonate of ammonia in doses of one drachm, or sweet spirits of nitre or sulphuric ether in doses of half to one ounce, repeated every four or six hours, until signs of relief are apparent. Steam the head well, as directed in " Sore Throat," with the addition of chloroform or chloric ether, as there stated. No bleeding or strong purgatives to be attempted ; but if purgatives are required, give Epsom salts, as directed in " Sore Throat." Hand rub the legs, and apply warm bandages to them ; clothe warmly. Rub fore part or sides of the chest with mustard or mustard and ammonia, for preparation of which see " Sore Throat." In about fifteen minutes after the application of the mustard, wash it off, and in two hours repeat, and again wash off, doing same at intervals until signs of relief are apparent.

INFLUENZA.

Causes.—Atmospheric, aggravated by close, ill-ventilated stables. Disease generally epizootic.

Symptoms.—Occurs chiefly in the spring; dulness and weakness; the animal is easily sweated; the droppings paler than usual, scanty, and partially coated with slimy matter; urine scanty; mouth hot and dry; yellowish red tinge round the gums and inside the eyelids; the eyes weep freely; the throat is sore; breath short; legs swelled; pulse quick and oppressed; horse staggers in his walk; flanks tucked up. A discharge of purulent matter from the nose in the early stage is a good sign. An unfavourable turn is indicated by fits of shivering, sometimes followed by profuse perspiration.

Treatment.—Remove the animal from contact with others to an airy loose box. Good nursing is the main thing to depend upon, and in many cases this alone is sufficient; avoid bleeding and strong purgatives, or strong sedatives. If it is required to act on the bowels, give Epsom salts in doses of one and a half to two ounces for several days. To lower fever give acetate of ammonia in doses of four ounces, with an ounce of nitric ether once or twice a day; but if there is much prostration with the fever, give spirits of nitric ether in doses of one to two ounces, at intervals of about four hours; and if the prostration increases, a draught of two drachms of nitrate of potassa, with two drachms of gentian and two of ginger, must be added every twelve hours while the symptoms continue urgent. Or the draught may be made of two drachms of carbonate of ammonia, one drachm of camphor, and two of ginger. If the throat is sore and swallowing difficult, apply a mustard poultice, washing it off in fifteen minutes, and repeating in two hours, if necessary. (For preparation of poultice see "Sore Throat.") Keep the body and legs warm, and let fresh water be always within reach.

The first symptoms of amendment will be the animal lying comfortably down. Nurse carefully as in pneumonia, &c., which see. Horses recovering from influenza are often attacked with a skin disease, small flattened lumps, which usually disappear spontaneously in a short time.

PNEUMONIA, PLEURISY, AND PLEURO-PNEUMONIA.

These diseases are closely allied in their causes, nature, and treatment.

Pneumonia is inflammation of the substance of the lungs, and is frequently combined with bronchitis.

Pleurisy is inflammation of the membrane which forms the covering of the lungs and also lines the cavity of the chest.

Pleuro-pneumonia is inflammation affecting both the lungs and the membrane which covers them and lines the chest. The disease may attack one lung or one portion of one lung, but it usually attacks both lungs at once.

Causes.—The causes are the same which produce other diseases of the respiratory organs, as in the case of catarrh, bronchitis, &c.

Symptoms.—In *Pneumonia,* slight cold, sudden fits of shivering, followed by coldness of the ears and extremities, and a staring coat. The horse is uneasy, and turns his head round to his chest. Pulse oppressed and quick ; breathing laboured ; the horse stands persistently with his forelegs wide apart, and his elbows out. In the early stage the lining of the nose is paler than usual, but it afterwards becomes of a purple, and finally of a leaden colour. When cough is present it is sharp at first, but as the attack progresses it becomes of a subdued character. During the early stage, if the ear be applied to the chest a confused, humming noise is heard, accompanied with a harsh, dry murmur, which, as the disease progresses, will give way to a moist rattle. This stage will last from 24 to 48 hours, at the end of which time a decided change for better or worse will occur. A very unfavourable symptom is afforded by the discharge from the nose becoming of a brownish colour. Occasionally the patient dies from conges-

C

tion of the lungs about the fourth or fifth day, or even as early as the second day, before any of the later stages are reached.

In *Pleurisy* the earliest symptoms are loss of appetite, quick, short respiration, and a quickened pulse. Then there is a clear, sharp grunt, when the animal is disturbed, or turned round in the stall, or on the application of pressure to his side. The pulse rises from 60 to 80 or 100, and even higher, becomes hard and wiry, but not so full or oppressed as in pneumonia. The patient does not lie down, but often attempts to do so. There are patches of sweat on the skin over the seat of the disease, and the muscles of the part are affected with twitchings. The inside of the nostrils becomes of a deep red colour. A marked sign of the disease is a regular elevated line or ridge along the lower border of the ribs, extending from the point of the hip to the lower part of the breast. If the ear be now applied to the chest a friction sound may be detected, similar to that caused by gently rubbing the dry hands together. There is great irregularity in the temperature of the extremities, portions of which may be very cold, whilst other portions are hot, and frequent alternations of temperature occur in the same part. Improvement is indicated by the breathing becoming less hurried, by the pulse becoming softer and more distinct, the cough less frequent, and the extremities continuing warm.

The symptoms in *Pleuro-pneumonia* are in the early stage, those of pneumonia, with the addition of the friction sound and elevated ridge across the ribs, noted as characteristics of pleurisy. The pulse is more affected than in pneumonia and less so than in pleurisy, and ranges about 70. The other symptoms are a combination of those of both pneumonia and pleurisy.

Treatment.—When the premonitory symptoms, such as slight cold, feverishness, dulness, or loss of appetite appear, give at once fresh air in a cool loose box; laxative diet, entire rest, extra clothing, and warm bandages to the legs. Abstain from giving corn or hay. Give diffusible stimulants, such as carbonate of ammonia, in doses of one drachm, or sweet spirits of nitre or sulphuric ether in doses of half to one ounce, repeated every four or six hours, until signs of relief are apparent, or the disease becomes worse.

If the fever be high, *but not otherwise*, during the first or dry stage, from five to ten drops of Fleming's tincture of aconite may be given, until relief is indicated by the pulse becoming softer and lower. In order to relieve the breathing, give two ounces of sulphate of soda or one ounce of nitrate of potassa, dissolved in a pailful of water, and the patient may be allowed to drink as much as he pleases. If he finishes the pailful give another, prepared in a similar manner. If the bowels are constipated give two ounces of Epsom salts dissolved in water, with half an ounce of nitrate of potassa twice a day. Apply friction and bandages to the legs, and if they still remain cold, apply a mustard plaister to them; wash it off after fifteen minutes, and rub with turpentine liniment, which is made as follows:—Take soft soap, 2 oz.; camphor, 1 oz.; oil of turpentine, 16 fluid ozs.; dissolve the camphor in the oil of turpentine, then add the soap, rubbing them together until they are thoroughly mixed. Re-apply the bandages. Apply mustard and ammonia—see "Sore Throat"—to the sides and chest; wash off after fifteen minutes, and repeat after two hours. Attend to diet, giving soft, nutritious food, attentive nursing, and extra warm clothing.

Bleeding in these diseases leaves the system unduly weakened, and violent blisters are apt to aggravate the fever and cause an unfavourable termination.

In the event of recovery, along with good nursing, fresh air, &c., some tonic medicines may be given, such as iodide of iron in doses of half a drachm twice a day, or sulphate of iron or of copper in half drachm doses, with two drachms of extract of gentian. Change or stop the tonic after a few doses. Mild medicine may be required to regulate the bowels. Bruised oats may be allowed in small quantities, along with fresh grass, carrots cut lengthways, bran mash, mixed with linseed, ground malt, skim milk, &c.

When these diseases terminate unfavourably, there is usually water on the chest, or an extensive adhesion of the lungs to the ribs, or the formation of abscesses, and sometimes mortification of the parts attacked.

/ The earlier symptoms of *water on the chest* may lead some to think that the animal is getting better. The extremities become warm, the pulse less frequent and soft, and the appetite partially returns; *but the patient still stands persistently with his forelegs wide apart.* A dropsical swelling appears between the forelegs, under the breast; the ridge along the belly becomes more distinct, and a straw-coloured discharge trickles from the nostrils. As the weakness increases the mane and tail become very loose, and may be easily detached. Small doses of one scruple of iodide of potassium, with gentian and ginger, may be given three times a day, with the view of stimulating the absorbents, and thus assisting nature in absorbing the effusion and deposit; but when the affection proceeds to these later stages, recovery is extremely doubtful, more especially if the vital powers have been weakened by bleeding and blistering.

ROARING—HIGH BLOWING—WHISTLING.

Causes.—Frequently caused in carriage horses by tight-bearing reins; also follows attacks of cold or influenza. It may be caused by a tumour in the nose, which must be removed by an operation.

Symptoms.—The obstruction to the passage of the air to and from the lungs causes the well-known noise which gives a name to this disease.

Treatment.—Hard food, with regular work, as in " Broken Wind" (which see), will exercise a favourable influence for a time ; but notwithstanding every care, the disease generally increases, until the animal becomes useless. If it is suspected to arise from tight reining up, remove the cause, and apply a blister or seton to the upper part of the throat. If it follows cough or influenza, rub well in on the outside of the throat the ointment of biniodide of mercury, repeated at intervals, or a seton may be inserted on both sides of the upper part of the throat. A roarer should not be bred from.

High blowing, piping, and whistling are all modifications of the disease.

SORE THROAT.

Causes.—Humidity of the atmosphere; bad ventilation in stables; neglect in stable management; allowing a sweating horse to stand in a draught or in the sun; irritation arising from the accidental presence of some foreign body in the throat.

Symptoms.—Cough, and difficulty of swallowing solids and even liquids. The mouth is hot, the horse is disinclined to eat, or " quids" his hay; that is, lets the masticated hay fall out of his mouth. He only sips his water, or takes it in small mouthfuls. The least pressure on the gullet produces coughing. There is much slobbering from the mouth, and when the animal drinks, a portion of the water, and even part of the food, comes back through the nostrils. /When the disease is complicated with strangles, the breathing is often accompanied by a roaring noise.

Treatment.—Turn into a well-aired loose box; restrict diet to soft food, such as grass, carrots, bran mash, or linseed gruel: *no hay whatever.* Avoid all active purgative medicines; but if the bowels are constipated, give two ounces of Epsom salts morning and evening, for two or three days, in a pint of water, with two drachms of ginger. If the salts do not act on the bowels, it will do so on the kidneys. Mix a tablespoonful of common salt with the bran mash. Give small doses of half a drachm of belladonna and an ounce of nitre, dissolved in water, and poured gently down the throat. Apply a mustard, or a mustard and ammonia, poultice to the throat, and when the disease is complicated with strangles, persist in the use of warm poultices or fomentations to the throat, and open any tumours or abscesses as soon as they begin to point.

A mustard poultice is made as follows :—Take mustard, 2½ ounces ; linseed meal, 2½ ounces ; warm (not boiling) water, 10 fluid ounces ; mix the linseed meal gradually with the water, and add the mustard, with constant stirring.　For a mustard and ammonia poultice, take of mustard and solution of ammonia, of each a sufficiency to form a poultice.　Oil of turpentine is sometimes added to this poultice.

Steam the head with vapours arising from boiling water poured on hay in a bucket ; and if there is much irritability of the membrane of the throat, pour four ounces of chloroform or chloric ether on the hay.　The patient will inhale it along with the steam. Maintain the warmth of the body by clothing, and wrap the legs in flannel bandages.　Keep water always within reach.　If these remedies fail, apply a blister of biniodide of mercury over the throat.　In the advanced stages give sulphate of copper in doses of one drachm in a pint of water, or two drachms of sulphuric acid in a similar quantity of water, once or twice a day.　Great care is necessary in giving a drench in cases of sore throat, and the horse's head should be let down immediately, if there is any attempt at coughing.

The first sign of recovery is a slight mucous discharge from the nostrils, and the cough will become softer.

Great care must be observed in the after treatment to prevent the animal becoming " a roarer," and it will be advisable to apply a strong blister of biniodide of mercury to the upper part of the throat.　Keep the horse in a cool loose box, avoid fast work for a time, and attend to the ventilation of the stable.

Gamgee gives the following liniment to be used in sore throat and other inflammatory diseases :—Soap liniment, 2 ozs. ; compound camphor liniment, 2 ozs. ; tincture of opium, ½ oz. ; mix.

STRANGLES

Cause.—Not clearly made out ; usually occurs in young horses from three to four or five years old.

Symptoms.—Horse sick and off his feed, with perhaps slight cold, and feverish symptoms. After a day or two the neck becomes stiff, and a swelling appears between the jaws. The enlargement at first is hard, hot, and tender; a discharge from the nose comes on; the symptoms increase, and the throat becomes sore; breathing is oppressed; the coat stares; the tumour softens, and on being opened the animal speedily recovers.

Treatment.—Avoid purging and bleeding. Keep up the strength of the colt by means of such soft, nourishing food as he can be got to swallow: a little fresh grass, or carrots cut lengthways, linseed gruel, oatmeal gruel, bran mash. Put some hay in a bucket, pour boiling water over it to steam the head, and if he picks afterwards a bit of the softened hay, well and good. Apply warm clothing to the body, and bandages to the legs; if the legs get cold, remove the bandages and rub the legs well. Keep the animal in a well-ventilated box. If the bowels are constipated give half a pint of castor oil, and repeat after 24 hours if necessary. Avoid all other purgatives, and if required resort to injections of warm water with soap or linseed oil mixed with it; keep the swelling warm with layers of flannel; and if assistance is necessary to bring it to maturity, apply a poultice of boiled carrots or turnips, but do not foment the part. The throat may be gently stimulated by using the following liniment:—Spirits of turpentine, 2 parts; laudanum, 1 part; spirits of camphor, 1 part; mix, and apply with a paste brush, morning, noon, and night, until the throat is sore. After each application cover with three pieces of flannel, and bind these on with an eight-tailed bandage.* When the tumour points externally, it should be opened as soon as it is nearly ready to burst. Keep the

---

* An eight-tailed bandage is a long piece of strong flannel, having three long slits made in each end, so as to leave four tails on each. The bandage is passed under the throat and jaw, and fastened by two of the tails from each end being brought over the forehead and tied, and two behind the ears.

incision open by putting a small piece of tow into the opening, and remove it occasionally. When the abscess is deep seated there is some danger in operating, unless the case is confided to a competent professional man.

The after treatment consists in good nursing and careful attention to diet and ventilation.

THICK WIND.

Causes.—Injudicious and violent exercise after watering, or when the stomach is full, or when the animal has been long kept on soft food. Pampered horses of a plethoric habit are liable to it.

Symptoms.—Owing to the air passages having become diminished in capacity, the horse cannot inspire the same quantity of air at each respiration, and is obliged, therefore, to breathe much more frequently than usual.

Treatment.—Incurable; and treatment can only be palliative. Good condition, regular work, and very careful watering and feeding will mitigate the evil. When the disease arises from a plethoric state of the body, purgative or diuretic medicines will be useful. Exercise and an occasional sweat will also be needed when it arises from over pampering and too little work. (See " Broken Wind.")

HEART DISEASE.

Cause.—Over exertion ; also exists in connection with other diseases.

Symptoms.—Heart throbs ; on moving in stall something like cramp noticeable in one of the hind legs ; when on a journey or hunting stops short and falls down ; in some cases on applying the hand to the left side a peculiar pulsation is felt, and on applying the ear to the same place the pulsatory sound is readily heard.

Treatment.—Death from disease of the heart is always sudden. If noticed in time, abstracting a little blood may be of use, and

also the administration of the following draught, which may be repeated if necessary :—Spirit of nitrous ether, 2 ounces ; tincture
of opium, 1 ounce ; tepid water, 12 ounces. Regulate the bowels
by means of a mild laxative, such as castor oil, and bran mashes,
and allow perfect quiet and rest for several days, avoiding severe
exertion of any kind after the horse is put to work. We have
met with a case of heart disease which, previous to death, was
supposed to be owing to a tight collar ; but examination after
death showed a fatty enlargement of the heart.

CHAPTER V.

THE STOMACH, LIVER, BOWELS, BLADDER, AND KIDNEYS.

BOTS.

Cause.—Bots are the grubs of the gadfly, the eggs of which
are deposited on the legs, arms, knees, or body of the horse during autumn, and are licked of by the animal and hatched in the
stomach.

Symptoms.—About June or July in the following year the bots
are voided in the dung, and when thus voided, often adhere to the
fundament.

Treatment.—As their presence does not usually act injuriously
on the health of the horse, there is no necessity for any special
treatment, as nature will quickly expel them without our aid. A

dose of physic may, perhaps, hasten the loosening of their hold on the coat of the stomach ; but we cannot force them to quit it much before nature disposes them to do so.

COLIC.

Causes.—Fast eating ; a surfeit of food after a long and fatiguing fast ; over drinking, especially when the body is warm ; sudden change of food ; bad food ; putting the animal to hard work on a full stomach.

Symptoms.—Sudden pain ; horse looks anxiously round to his flanks ; knocks about, lies down and gets up frequently or rolls over and kicks. Fits pass away, but soon return. Pulse quick ; dung pellets hard and angular ; belly tense, and sometimes swollen, and very tender on pressure.

Treatment.—Give linseed* or castor oil, a pint, and an ounce of nitric or sulphuric ether ; where the ether cannot be had, substitute gin, rum, or whiskey, with some pepper and ginger. The belly must be well rubbed, and a stimulating embrocation of turpentine may also be rubbed on the belly at same time. For this, take oil of turpentine, 3 ounces ; olive oil, 6 ounces ; mix. Dissolve a pint of turpentine in a quart of solution of soap, and give it as an injection. If relief is not obtained in half an hour, give another ounce of sulphuric ether and eight to twelve ounces of castor oil. Continue rubbing the belly, and give some gentle walking exercise at intervals of half an hour, and if the spasmodic attacks still continue at the end of an hour and a half or two hours, apply hot fomentations by means of a rug steeped in hot water and held to the belly by a man on each side. To prevent chill when the fomentation is discontinued, rub the belly with ammonia liniment ; that is, solution of ammonia, one ounce ; olive oil, two ounces ;

---

* Linseed oil, when pure, is a safe and effective purgative, but the linseeds, especially some foreign kinds, are so often mixed with those of other plants possessing an acrid quality, that the linseed oil now obtained has become both uncertain and unsafe in its action.

mix, and shake well before using. Perseverance in these remedies will generally be sufficient to overcome the attack. After an
attack of colic, prepare the horse for physic ; that is, let him be
deprived of all food for 36 or 48 hours, except cold bran mashes,
and this of itself will generally be all that is required, without
actually giving a dose.

Farm horses are very liable to colic, owing chiefly to the careless manner in which they are frequently fed, watered, and
washed. Professor Dick recommends the following tincture to
be kept in readiness :—Whiskey or brandy, 2 pints ; Cayenne
pepper, 1 ounce ; ginger, 3 ounces ; cloves, 3 ounces ; digest for
eight days, and then add sweet spirits of nitre, 4 ounces. Half a
pint of this tincture is a dose, in a quart of warm water.

INFLAMMATION OF THE BOWELS.

Causes.— Follows a prolonged attack of colic ; continued
constipation ; indigestion ; excessive action of a purgative ; over-
hard work ; exposure to cold when the animal is sweating.

Symptoms.—The early symptoms are the same as those of colic ;
but with this distinction, that the pain is continuous, whereas in
colic it is intermittent. Pulse very high ; extremities cold ; mouth
dry and either unnaturally hot or cold ; respiration hurried and
oppressed ; nostrils much dilated ; countenance anxious ; body
bathed in sweat and then cold, with occasional tremors ; tail
erect and quivering.

Treatment.—Bleed copiously, unless there is great prostration ;
give an ounce of tincture of opium, and the same of spirit of
nitrous ether, followed by doses of 12 ozs. of linseed or castor oil
and half a drachm of opium every three or four hours. Apply hot
fomentations to the belly as in "Colic ;" also mustard freely to the
same place. Give frequent injections of small quantities of warm
water, but avoid walking exercise. Should the animal recover, great
care and attention will be required as regards the diet, which
should consist of soft food of an easily digested character.

STRANGULATION AND RUPTURE OF THE BOWELS.

Causes.—Obscure; but sudden exertion with an over-loaded stomach is dangerous.

Symptoms.—Resemble those of colic, or rather of inflammation of the bowels; rapid sinking of the system; pulse weak and fluttering; sometimes the horse sits on his haunches.

Treatment.—Useless; although, of course, it is probable that until the actual nature of the disease is known, similar treatment to that adopted in inflammation of the bowels will likely be followed.

CALCAREOUS AND DUST BALLS.

These bodies are found in the stomach and large intestines of the horse, and vary in size from that of a marble to several pounds in weight.

Causes.—Calcareous balls are ascribed to the calcareous character of the district or of the water drunk. Dust balls are most common in millers' horses, and arise from the liberal use as food of the dust of corn and barley saved in grinding.

Symptoms.—Frequent colicky pains, which pass off without medicine; the animal evidently suffers in health; the countenance is haggard, the eye distressed, the back up, the belly distended; the respiration becomes hurried; bowels habitually costive; and the horse when attacked with pain from this cause will sit like a dog on his haunches.

Treatment.—Strong purgatives, and large injections of warm water with soap dissolved in it.

DIARRHŒA.

This disease is most frequently witnessed in what are called "washy" animals.

Causes.—Change of food, such as fresh grass, new oats, and hay; irritation in the bowels; chill after over exertion.

Symptoms.—The copious and frequent evacuation of the dung in a watery state.

Treatment.—In slight cases starch gruel, or gruel made of wheat flour, with some chalk, and powdered opium; one drachm of the latter for a dose will be sufficient. Change the food. If there is irritation, give a laxative medicine, such as a pint of castor oil, and two to four drachms of powdered ginger in gruel. In ordinary cases give the following in gruel made with flour :—

Powdered ginger	... ...	1 drachm.
„ gentian	... ...	2 drachms.
„ opium	... ...	$\frac{1}{2}$ drachm.
Prepared chalk	... ...	1 ounce.

Mix, and repeat twice or thrice a day. This will also be given after the oil has cleared out the bowels.

When foals are attacked with diarrhœa give, in flour gruel— Prepared opium, 1 scruple; powdered rhubarb, $\frac{1}{2}$ ounce; powdered gentian, $\frac{1}{2}$ ounce; prepared chalk, 1 ounce; to be well intermixed.

DYSENTERY.

Cause.—Irritation of the bowels from some acrid substance.

Symptoms.—The dung is passed in a half solid form, covered with slimy matter; and emits an offensive smell. There is abdominal pain, perspiration, and great internal heat, which may be detected by placing the hand up the fundament.

Treatment.—Give the following drink :—

Sulphuric ether	... ...	1 ounce.
Laudanum	... ...	3 ounces.
Liquor potassæ	.. ..	$\frac{1}{2}$ ounce.
Powdered chalk	... ...	1 ounce.
Tincture of catechu	... ...	1 ounce.
Cold linseed tea	... ...	1 pint.

Mix, and repeat every fifteen minutes, along with injections of cold linseed tea, until the irritating matter has been expelled.

Cleanse the quarters, and plait the tails. After recovery, the food for a week must consist of linseed tea, hay tea, and gruel. On the expiration of a week, a few boiled roots may be added; three of the doses above recommended being given every day for a fortnight. Then give some scalded oats, but no cold water or hay for a month.

DROPSY OF THE ABDOMEN.

Symptoms.—Pulse hard; head hangs down; mouth dry; pressure on the belly elicits a groan; turning in the stall a grunt; listlessness; frequently lying down; restlessness; thirst; loss of appetite; weakness; thinness; enlarged belly; constipation and hide-bound. Small bags hang from the chest and belly; the sheath and one leg sometimes enlarge; the mane breaks off; the tail drops out. Purgation and death.

Treatment.—Give the following, night and morning:—Strychnia, half a grain, which may be gradually increased to one grain; iodide of iron, half a drachm, increased gradually to one drachm and a half; extract of belladonna, one scruple; extract of gentian and powdered quassia, of each a sufficiency; mix. Apply small blisters, in rapid succession, upon the belly; but if the effusion is confirmed, a cure is hopeless. Professor Dick recommends iodine, in doses of one or two drachms, twice a day, as having an excellent effect.

HERNIA, OR RUPTURE.

This disease rarely occurs in any but entire horses.

Causes.—It is sometimes congenital, or born with the animals; violent exertion will also produce it.

Symptoms.—Scrotal hernia consists of a protrusion of a portion of the gut into the scrotum. It may exist for a time without injury; but there is a constant liability to strangulation, when symptoms of acute pain similar to colic are manifested.

Treatment.—Some kinds of hernia are invariably fatal, but

scrotal hernia is generally easily rednced, if detected early, and before strangulation occurs. Gamgee says, attempt reduction by manipulation, and if unsnccessful aloue, try warm fomentations or warm bath, ice, and the nse of chloroform. If this be unsuccessful, relieve straugulatiou by dividing the stricture. After the operation, administer a purgative, and keep the animal quiet. It is evident, however, that it is only a professional veterinary surgeon who can treat hernia, and no other person should attempt it.

INFLAMMATION OF THE LIVER.

Causes.—High living, and too little work.

Symptoms.—Cold mouth; white of the eyes displays a yellow tinge; dung usually hard and coated; right side tender.

Treatment.—Moderate bleeding in the first instance, also a blister on the right side. Open the bowels moderately with linseed oil. Then give the following ball, twice a day, for several days:—

Calomel	.	.	1 drachm.
Opinm	.	.	1 scruple.
Nitrate of potash	.	.	2 drachms.

Mix. Mayhew recommends a sufficiency of nutritious food, but only enongh of it; plenty of labour, and the following physic:—

Iodide of potassium	.	2 ounces.
Liquor of potassæ	.	1 quart.

Mix. Dose, night and morning, two tablespoonfuls in a piut of water.

WORMS.

Cause.—Poverty in the system.

Symptoms.—The coat has a peculiar hard, dry, and nnthrifty appearance; appetite ravenous; body thin; bad smell from breath; the animal picks and bites its own hair; rubs its nose against the wall; and when the worms exist in the rectum, it ruhs its tail and hind-quarters agaiust the wall until the hair is rubbed off.

Treatment.—Spirits of turpentine in linseed oil will generally give relief. To a foal give 2 drachms of turpentine; to a six months old, 1 ounce; to a year old, 1½ ounce; to a two year old, 2 ounces; and so on up to 4 ounces, the quantity of oil varying from an ounce to a pint, according to age. The after treatment to restore lost condition consists in giving the following ball, daily, for a week:—

Sulphate of iron	. .	1 drachm.
Powdered gentian	. .	2 do.
Powdered ginger	. .	1 do.
Powdered pimento	. .	1 do.

Make into a ball with treacle. Give rock salt in the manger.

INFLAMMATION OF THE NECK OF THE BLADDER.

Symptoms.—Retention of the urine; uneasiness; colicky pains; straining to void urine; clammy sweats.

Treatment.—If caused by accumulation of dirt about the opening of the urethra or urine passage, a thorough washing will often be sufficient. Give injections of tepid water; pass the hand through the anus to the bladder, and press on the neck of it; but if these means fail to cause a discharge of the urine, a catheter must be used. This is easily done in the mare, but in the horse the operation requires care and skill. The penis being firmly grasped with the left hand, and drawn out from the sheath, the catheter, previously oiled, is gently and steadily pushed up the urine passage until it enters the bladder, when the whalebone stilette may be withdrawn from the tube, so as to permit the urine to flow. After relief, attend to the nursing and diet of the animal, and avoid for some time exposure or over-hard work.

BLOODY URINE.

Causes.—A diseased state of the kidneys or other parts of the passage; also violent strains and internal ruptures.

Treatment.—Rest and laxative diet, especially grass; also

linseed tea. If the disease arises from any weakness of the urinary organs, give half a drachm each of nitric and muriatic acid, with two ounces of gentian, daily. Covering stallions sometimes suffer from bloody urine, brought on by the excessive use of stimulants given to them by their caretakers. In such cases avoid the cause.

CALCULI, OR STONE IN THE BLADDER.

Cause.—Not well ascertained; probably calcareous nature of the water used as drink.

Symptoms.—Urine thick, dark or bloody, and passed in a dribbling manner; back roached.

Treatment.—Hydrochloric acid given twice a day in doses of two drachms in a pint of water will, in some cases, when the calculus is small, and no urgent symptoms present, be successful in dissolving the concretions; but if the calculus is large and symptoms severe, an operation must be performed, which can only be done by an experienced veterinary surgeon.

DIABETES, OR EXCESSIVE STALING.

Causes.—Bad forage, such as mow-burnt or mouldy hay, and kiln-dried oats. It also follows the frequent use of "condition balls," which are chiefly composed of nitre or turpentine.

Symptoms.—Excessive staling, accompanied by extreme thirst; skin dry, coat rough and staring; digestion out of order; bowels torpid; appetite capricious; horse sweats easily, and falls away rapidly in condition.

Treatment.—If arising from bad forage, change it; give grass, carrots, laxative diet, linseed tea in place of water, and a slight dose of linseed oil. If the water is hard, let it be boiled before being used as drink. Give iodide of iron in doses of one drachm daily; or of iodide of potassium, one drachm; or iodine, half a drachm daily. Feed on sweet, fresh oats, pale malt, not high

D

dried, and boiled peas. Clay mixed with the water given to drink sometimes acts beneficially.

INFLAMMATION OF THE KIDNEYS.

Causes.—Prolonged or severe work ; exposure to wet or cold ; excessive use of diuretic medicines ; bad provender, such as mow-burnt hay, kiln-dried oats, or other irritating food. It may also arise from the pressure of calcareous matter in the kidneys.

Symptoms.—Feverishness ; restlessness ; free perspiration ; lies down and rises up as in colic, but the belly is tucked up, instead of being distended, as in that disease. The horse is unwilling to move ; stands with his legs wide apart ; crouches and straddles in his gait, and groans if turned sharply round. Tenderness evinced by wincing when pressure is applied to the loins. At frequent intervals he appears about to stale, but passes no urine, or only in small quantity, highly coloured, and often tinged with blood. In very severe attacks the horse will sit on his haunches, groan, and look round to his flanks. The bowels are usually constipated ; pulse quick and weak. When the inflammation is caused by calcareous substances in the kidneys, the penis hangs pendulous, and there is a constant dripping of urine, often tinged with blood. In the case of a stallion, the testicles are retracted. The thigh of the side of the inflamed kidney, if only one be affected, is generally benumbed. Unless relief is given, the strength fails rapidly, and the skin often acquires a urinous smell.

Treatment.—Apply flannel cloths steeped in very warm water to the loins. Give a pint of linseed or castor oil with a scruple of calomel. After 48 hours the oil may be repeated if necessary. When the purgation has ceased, the calomel may be repeated in half-drachm doses with one drachm of opium in a ball, night and morning, for three or four successive days. Aloes must not be given, nor saline purgatives. If the pain and straining is great, give belladonna in doses of one drachm, combined with one drachm of opium, twice a day, for two days; but not more. Lay the skin of a newly-flayed sheep across the loins, with the flesh side inwards, and change for a fresh one every second day. When the sheepskin or fomenting cloths are removed, in order to prevent chill, rub the parts well with an embrocation wash of six ounces of linseed oil, one ounce of the strongest water of ammonia, and two ounces of tincture of opium. Injections of warm water are useful. Give linseed tea, hay tea, carrots, grass; but not clover or vetches. Let the animal be supplied with tepid water, slightly acidulated with sulphuric acid. When the urine begins to come away, give half an ounce of bi-carbonate of soda, two or three times a day, to lessen the acidity of the urine; and if there is still much pain, give a ball composed of a drachm of powdered opium with half a drachm of camphor.

After an attack of this kind, the kidneys will always continue irritable, and the greatest care will be required afterwards as regards food and general treatment.

CHAPTER VI.

THE SKIN.

SURFEIT OR NETTLE-RASH.

Cause.—Heat of body, arising from plethora, which is not exactly fatness, but the formation of more blood than the system can easily dispose of, and it becomes oppressed. A sudden change from poor to rich keep will cause plethora.

Symptoms.—The skin is raw, and covered with a number of round swellings about the size of a shilling, which appear all over the body, but particularly about the neck and quarters; or the eruption may be smaller, and more like pimples. The hair falls off in patches.

Treatment.—Give a gentle purgative ball on an empty stomach after a course of cold bran mashes, and follow up with the following medicine, after the purgative has ceased to operate:—Yellow sulphur, 4 drachms; nitrate of potash, 3 drachms; antimony, 2 drachms; mix, and give in the food twice a day. Mayhew recommends the following alterative and tonic drink:—Liquor arsenicalis, one ounce; tincture of muriate of iron, one ounce and a half; water, one quart; mix; give daily, one pint for a dose. Let the food consist chiefly of grass, carrots, bran mashes, and add a handful of old beans to the oats. Keep the stable or box well aired and clean.

LICE.

Causes.—Neglect of grooming; filth; poverty; mange; hidebound, &c.

Treatment.—Rub the skin, and particularly the parts affected, with olive oil. If this is not effective after repeated use, apply some of the mixtures recommended for mange, such as carbolic acid, &c.

MANGE.

This is a most disagreeable disease, and so highly contagious that it will be communicated not alone by actual contact, but also by standing in the same stall, by using the same harness and stable requisites, and even by the hands of the attendants.

Causes.—The attack of a minute insect, similar to that which produces itch in man, which burrows under the skin, and breeds rapidly. These insects are never found in healthy, well-groomed, well-fed horses; and Col. Fitzwygram judiciously remarks that

" the owner will do well to change his servants whenever this disease appears in his stable."

Symptoms.—Mange usually begins at the roots of the hair of the mane and tail. A number of small scabs appear, abounding on the withers and buttocks, and spread over the body. The hair becomes rubbed off the body, from the horse rubbing violently against the stall, and biting his skin wherever he can reach it. The skin of the neck and sides becomes thickened, and formed in hard, dry folds. The itching appears intolerable.

Treatment.—Brush the skin thoroughly, and then dress the parts affected with a solution of carbolic acid in the proportion of half an ounce of the acid to a pint of water. One dressing will be sufficient, if the disease is taken in time. About 48 hours after application it may be washed off.

Or, the skin, having been first thoroughly washed with soft soap and warm water, with soda dissolved in it, a liniment consisting of equal parts of oil of tar, oil of turpentine, and common oil may be applied, by means of a brush, every second day for a week, and then washed off with soft soap and water.

Or, after washing the skin as above, rub in with much friction the following ointment :—Flowers of sulphur, 4 ounces ; white hellebore, 2 drachms ; oil of tar, 4 ounces ; linseed oil, 1 pound ; to be mixed together. Repeat daily for several days ; then wash off with soap and water ; repeating dressing, if necessary.

Give green food, pale malt, and boiled linseed, or oil-cake, if green food is not procurable. Cleanse thoroughly the stalls, washing with diluted carbolic acid, and boil or bake the clothing.

For ordinary itchiness, not mange, but owing to which patches of hair are sometimes rubbed off ; or for an affection of the skin which sometimes appears about the head and face, or under the brow-band, dress the parts affected with equal parts of mercurial ointment and soft soap made into a lather with hot water, and applied by means of an old brush. Change the diet, and give laxative food as above.

WARTS.

Cause.—Obscure.

Symptoms.—Warts usually appear on the thin and more delicate portions of the skin; as, for instance, the sheath, the inner side of the hind limbs, the belly, the eyelids, and the sides of the nose. Occasionally they are found on the neck, when the skin has been injured by the collar.

Treatment.—Scrape the surface and then dress it with chloride of zinc. Large warts may, however, require to be removed by the knife, followed by the application of the hot iron to stop the bleeding. Sometimes the wart has a neck or small base, and in that case tie a waxed silk thread tightly round the neck, and after a time the wart will drop off. The skin over small warts may also be slit open and the wart squeezed out by the fingers. Avoid all arsenical applications to warts.

Tumours are sometimes mistaken for warts; and it is only a practical veterinary surgeon who can deal with such, as they may require to be cut out, and are sometimes connected with other serious maladies.

WARBLES

Causes.—The larva of the gadfly, produced from the eggs laid in the hair during the previous season's " run at grass."

Symptoms.—Small, soft tumours in the skin.

Treatment.—Open the abscess with a lancet, and squeese out the larva. Apply to the wound a lotion made of chloride of zinc, one grain; water, one ounce. Dab this on with a linen rag.

HIDE-BOUND.

Cause.—Either the accompaniment of a disease or the result of poverty.

Treatment.—Good food of a laxative nature, clean stable, pure air in abundance, salt sprinkled on corn, healthy exercise, and good grooming.

MALLENDERS AND SALLENDERS.

Cause.—Filth and neglect.

Symptoms.—In mallenders there is a scurfy and somewhat obstinate eruption on the back of the knee in the fore-leg; and in sallenders, a similar affection in front of the hock in the hind one.

Treatment.—Cleanliness; laxative diet; rub in well and frequently the ointment of carbolic acid, or the following ointment:—

Animal glycerine	... ...	1 ounce.
Mercurial ointment	... ...	2 drachms.
Powdered camphor	... ...	2 do.
Spermaceti	... ...	1 ounce.

Mix thoroughly. If cracks appear, treat as if for cracked heels.

SITFASTS.

Causes.—Badly fitting saddles or collars; bad riding; neglected warbles.

Symptoms.—Hard, bare tumours, like a corn on the human foot, but frequently surrounded with a circle of ulceration.

Treatment.—Apply Goulard water to disperse the swelling, and a digestive ointment; such as—oil of turpentine, 8 ounces; camphor, 1 ounce; soft soap, 4 ounces; to be shaken together till mixed. The sore should be healed with a solution of sulphate of zinc. It will sometimes be necessary to cut out the "sitfast," in which case lunar caustic should be applied to the part; or the solution of sulphate of zinc or of chloride of zinc.

SADDLE-GALLS.

Cause.—Friction of the saddle.

Treatment.—Apply a strong solution of salt with tincture of myrrh, and attend to the padding of the saddle. The same applies to shoulder-galls, arising from ill-fitting collars.

A veterinary surgeon, George H. Dadd, gives, in the *Prairie*

Farmer, the following treatment of " sitfasts" and " saddle-galls :"
—So soon as an abrasion is discovered on the back of a horse
the animal should be excused. from duty for a few days ; the
abraded parts should be dressed twice daily with a portion of the
tincture of aloes and myrrh. This simple treatment will soon
heal the parts. Should there be no abrasion, but simple swelling,
attended with heat, pain, and tenderness, the parts should be
frequently sponged with cold water. Occasionally the skin under-
goes the process of hardening (induration). This is a condition
of the parts known to the farriers of old as "sitfast," and the
treatment is as follows: Procure one ounce of iodine and smear
the indurated spot with a portion of the same twice daily. Some
cases of galled back and shoulders are due to negligence and
abuse ; yet many animals, owing to a peculiarity of constitution,
will " chafe," as the saying is, in those parts which come in contact with the collar and saddle, and neither human foresight nor
mechanical means can prevent the same.

RINGWORM.

Cause.—An unhealthy condition of the skin, produced by neglect
of grooming, bad food, or sudden change of diet. Damaged oats
or bad hay are said to be very ready causes of this disease.

Symptoms.—Coat unthrifty ; hair falls off in patches, exposing a
scurfy skin. Blotches are formed, which assume the form of rings.

Treatment.—Clip off the hair round the blotches, and wash
the parts well with soap and water, after which a strong solution
of sulphate of iron or copper should be applied two or three days
in succession.

The following dressings have also been much recommended :—
Mercurial ointment; biniodide of mercury ; also oxide of zinc
ointment ; or a mixture of half an ounce each of sulphur, iodide of
potassium, and iodine, made into a liniment with eight ounces of
lard.

Change the food ; give a bran mash with a pound of boiled

linseed daily ; and an alterative, consisting of two drachms of nitre and half a drachm of sulphur, may be given for two or three days.

ANNOYANCE FROM FLIES.

Although this is not a disease, yet as being the cause of much discomfort in riding and driving, we give the directions for prevention, as recommended by a writer in the *Journal of Chemistry.* Take two or three small handfuls of walnut leaves, upon which pour two or three quarts of cold water, let it infuse one night, and pour the whole next morning into a kettle, and let it boil for a quarter of an hour. When cold it is fit for use. No more is required than to moisten a sponge, and before the horse goes out of the stable, let those parts which are most irritable be smeared over with the liquor, namely, between and upon the ears, the neck, the flanks, &c. Not only the gentleman or lady who rides out for pleasure will derive pleasure from the walnut leaves thus prepared, but the coachman, the waggoner, and all others who use horses during the hot months.

CHAPTER VII.

THE LIMBS AND FEET.

BOG SPAVIN.

Cause.—Believed to arise from hard usage of some kind.

Symptom.—A soft, elastic, puffy swelling at the front of and at the upper part of the hock.

Treatment.—Pressure, maintained by means of an India-rubber bandage. If this fails, apply repeatedly with friction the ointment of the biniodide of mercury, and when soreness has been induced by it, abstain from the use of it for a time ; but as soon as these effects have passed off it may be again and again resorted to.

CRACKED HEELS.

Causes.—Washing the legs with warm water, and allowing them to dry by evaporation ; neglecting to dry after washing with cold water, after exercise ; clipping the hair off the back part of the fetlock, especially in draught animals ; dirty-kept stables.

Symptoms.—The hind heels more often affected than those of the foreleg ; disease more common in winter than in summer The heel becomes hot, tender, and swollen, and the skin cracks in various places. On first coming out of the stable lameness will be noticed, but this goes off. The swelling in bad cases may extend to the part of the leg above the heel, and some blood may ooze from the cracks.

Treatment.—Wash the heels every morning with warm water, and then carefully envelope them in flannel bandages. Do not use soap with the water. If the cracks do not readily yield under this treatment, a weak solution of nitrate of silver may be applied, and it may be necessary to stimulate the parts with turpentine liniment, and to touch the cracks occasionally with sulphate of copper or nitrate of silver (lunar caustic). Poultices may be needed in severe cases, but warmth and pressure and stimulating lotions usually answer best.

GREASE.

Causes.—Same as in the case of " Cracked Heels."

Symptoms.—The first is an intolerable itching of the heels, indicated by frequent stamping and rubbing of one leg against the other, generally followed by swelling of one or both legs. In the next stage a discharge, like drops of oil, will be found clinging to the hairs at the hollow of the heel, and the hairs stand erect, so that the skin becomes visible. The heels feel hot and greasy. The horse moves stiffly, and it gives him great pain to lift his leg or to have it lifted. The swelling increases, and the animal becomes more and more painfully lame, and straddles in his walk. The discharge has an offensive smell; the heels become excessively sensitive, and if not checked the disease will run on to ulceration, and granulations, called " grapes," form in bunches, the points of which harden like horn; legs swell to twice or thrice their natural size, the hair falls off, and the horn of the hoof grows long.

Treatment.—The first thing is to remove the cause, and take care that it is not again allowed to occur. In simple cases apply a warm linseed meal poultice, changed twice a day, for two or three days. The discharge should be cleared away with a sponge and tepid water on each occasion before the poultice is renewed. Trim the hair off. Let the diet be of a laxative nature, and if the animal is not weak, give a dose of purgative medicine. When this has produced its effects, apply pledgets of tow saturated with equal parts of sulphate of zinc and sugar of lead dissolved in water. The pledgets must be secured by flannel bandages. If the discharge is of a virulent nature a stronger lotion will be required, such as an ounce of sulphate of copper dissolved in a pint of water. If the legs are much disposed to swell give a diuretic ball (see " Water Farcy") every other day. As soon as the discharge is arrested substitute an astringent ointment for the lotion, such as the compound ointment of alum ; or the following :—Tannin, 2 drachms ; glycerine, 2 ounces ; mix. The ointment of carbolic acid may also be used with advantage. In some cases it may be necessary to re-apply a poultice for a day or two. If ulcerations have formed, dress with the following preparation until a healthy action is, if possible, established :—Nitrate of silver, 8 grains ; glycerine, 2 ounces ; mix. This application also answers well in cases of inveterate cracked heels. When grapes are formed, try first the introduction of powdered sulphate of zinc among them, as this has often been found effectual, when properly done and persevered in, without any further application; but if unsuccessful, the bunches of proud flesh must be cut out with a keen knife, and cauterised with the red hot iron. For this end the animal must be cast, and it may be necessary to go over them more than once, according to the strength of the animal, so that it is only an experienced professional veterinary surgeon who should attempt it.

In recovery, the horse must be prevented, by means of a cradle, from biting the itchy parts. The food should consist of green meat, carrots, bran mashes; no corn or hay. Regular exercise is indispensable during recovery, but not during treatment, when the animal should be confined in a roomy, and very airy, loose box. Bandages and pressure hasten the cure.

SWOLLEN LEGS.

Causes.—Some degree of debility in the system.

Treatment.—Place in a loose box; hand rub the legs well and long; and give regular exercise. Give grass instead of hay; or carrots and bran mashes; damp the corn and sprinkle a little ground oak bark on each feed.

BREAKING DOWN.

Causes.—Strains of the flexor tendons, brought on by violent exertion in racing or leaping.

Symptoms.—The horse, when going, suddenly loses power to put one foot to the ground. The foot is turned upward; there is excessive pain; quickened breathing, and accelerated pulse; appetite gone.

Treatment.—When the fetlock does not wholly come to the ground, the foot may be rendered useful; but when both fetlocks come down, the horse will seldom recover. Bleed, and if necessary, purge to remove fever, and apply frequent warm fomentations to the parts, and in the intervals, cotton saturated with tepid water should be applied to the leg, and covered with oiled silk, to retain the moisture. Cold lotions may succeed, such as the following:—Muriate of ammonia, 4 drachms; nitrate of potash, 2 drachms; pyroligneous acid, 1 ounce; cold water, 1 pint. Mix. Put a patten or high-heeled shoe on the foot. When the inflammation subsides, the parts may be fired or blistered, or rubbed repeatedly with the ointment of biniodide of mercury. The accident sometimes leads to a contraction or drawing up of the leg, which knuckles over at the fetlock. For relief of this the tendons must be cut.

CANKER IN THE FOOT.

Causes.—Neglected thrush, sandcrack, or other diseases, and injuries of the foot, when not attended to.

Symptoms.—Not much lameness. The sensible frog and other parts, instead of secreting horn, produce a fungous growth, which pervades the whole sole, and ultimately extends to the entire secreting surface of the foot. Almost peculiar to heavy cart-horses, and sometimes it would seem there is a strong hereditary disposition to it. Horses with white feet are most liable to attack, and the hind feet more than the fore.

Treatment.—If it has existed for some time it is difficult to cure. Pare off the superficial fungoid horn, and as much of the deep-seated as can be detached. Stop bleeding by the hot iron. Professor Dick says that "M. Feron regarded tar and sulphuric acid, in the proportion of four ounces of the former to two drachms of the latter, as a specific." Cleanliness, perseverance, and time will effect a cure; but with a dressing of tar in which verdigris and nitric acid, two drachms of each to one pound of tar, are well mixed, and applied with a degree of *firm pressure*, at least every second day, the worst cases may be got under. Chloride of zinc, half an ounce; flour, four ounces; mixed and applied every second day to the diseased parts, is also beneficial.

CONTRACTION OF THE BACK SINEWS.

Causes.—Over exertion ; repeated strains.

Symptoms.—This occurs chiefly iu draught horses, engaged in drawing heavy loads. The sinews become shorter; the horse steps at first on his toe, and afterwards the front of the fetlock nearly, or quite, comes to the ground. In a short time a tumour appears, which at first is small, soft, hot, and tender, but soon grows hard. The animal suffers great pain.

Treatment.—Bleed and physic gently, and administer febrifuge medicines to reduce the pulse to 55.° The following are examples of febrifuge medicines :—

<pre>
        No. 1.—Camphor   ...   ...   1 drachm.
               Nitre   ...   ...   2 do.
                    Make into a ball.
        No. 2.—Nitre   ...   ...   2 drachms.
               Epsom salts   ...   1 to 4 do.
                 Dissolve in a pint of warm water.
</pre>

Put a linen bandage on the leg, and keep it constantly wet, covering it with oil silk until the primary symptoms abate. Give cut grass, aud cooling food generally, with a long rest in a loose box.

CORNS.

Causes.—Bruise of the sensible sole.

Symptoms.—The seat of the corn is at the inner angle of the sole, in the space between the crust and the bar. Sometimes matter forms, rises upwards, and makes its appearance between the hair and the hoof. Flat and contracted feet are particularly liable to corns.

Treatment.—Remove the shoe and pare away the part until the quick is nearly exposed. Put a linseed poultice on the foot, and continue it for several days if the case is severe, after which touch the part daily with the butyr of antimony ; or, in bad cases, apply a strong solution of sulphate of zinc. When a healthy surface has been obtained, the following ointment will greatly assist in promoting the growth of horn :—Oil of turpentine, 4 drachms; sulphuric acid, 4 drachms; to be carefully and gradually mixed together in an open place, and after the boiling has subsided, add Barbadoes tar, 8 ounces ; palm oil, 4 ounces ; and mix all well together. Tack on an old shoe ; give rest.

FALSE QUARTER.

Cause.—Injury to the coronet.

Symptoms.—No lameness, but weakness of the foot. Sometimes a cleft or fissure is formed, and sometimes the diseased part exhibited by a somewhat modified and inferior kind of horn. The parts are apt to bleed, and when granulations sprout there is much pain and lameness.

Treatment.—Same as in " Sand-crack."

FOUNDER, OR FEVER IN THE FEET.

Causes.—Great stress and exertion ; standing long on shipboard, where there is no room for movement ; it is also produced when the animal gets loose in the stable and gorges himself at the corn-bin.

Symptoms.—Appears more frequently in the fore than in the hind feet, but may attack all the four. Great pain, disinclination to move, and an unwillingness to throw the weight on the inflamed feet. The animal stands with his hind legs drawn up under his body. If compelled to move by pushing him backwards, he plants the heels of the forefeet on the ground, and brings the hind legs well forward. There is a great heat in the foot, and fever. The whole crust may become separated from the senseless foot, leaving the stump bare and exposed ; or if checked, the separation may be partial, or wholly absent, and there may be no greater mischief than, by-and-by, the appearance of a slight depression or ring upon the crust.

Treatment.—Sling the animal, remove the shoes, and set the feet in pails containing warm water, in which some caustic soda has been dissolved. Mayhew recommends, before the shoes are removed, to give half a drachm of belladonna and fifteen grains of digitalis every half hour until the symptoms abate. Bleed from the neck, clothe the body, give thin gruel and green food; and have two men to watch for the first three nights. Next day give sulphuric ether and laudanum, of each two ounces, in a pint of water. Should the pastern arteries throb, open the veins, and place the feet in warm water. While the affection lasts, pursue these measures; and it is a bad symptom, though not a certain one, if no change for the better takes place in five days.

NAVICULAR DISEASE—GROGGINESS.

Causes.—Hard roads and a fast pace, especially after long confinement in the stable or limited exercise on soft ground; treading on a rolling stone, or travelling with a stone in the foot; tight shoeing, and hereditary tendency in well-bred horses, to which class it is chiefly confined, being rarely found in agricultural horses.

Symptoms.—Lameness in one or both fore feet, which diminishes as the horse gets warm from exercise; the action is low and short; the animal goes greatly on his toes, and is, therefore, apt to trip and stumble; in the stable the affected foot is held out in a pointed direction, and if both are affected there is a shifting of the feet. The heels become contracted, and there is a degree of heat in the foot, more especially about the heel and coronet.

Treatment.—All possible attention must be paid to the shoeing; the sole should be thinned, the bars pretty freely removed, and the toe made short. The powers of the animal must not be overtasked. Blister the pasterns, and protect the parts by leather soles, and also by *stopping* by means of tow and moss wetted, or cow dung mixed with a little clay, or felt pads. Keep the animal in a well ventilated loose box; give gentle exercise, and with time and patience much may be effected. Should this treatment fail, recourse must be had to the operation of *neurotomy,* which only a practised professional man can perform.   That operation removes pain and lameness, but as sensation is entirely destroyed, no dependence can be placed afterwards on the animal ; and if pushed to over-exertion he may knock his hoofs off altogether.

PRICK OF THE SOLE.

Causes.—The entrance of one or more nails into the sensible parts within the hoof, or so close to them as to cause pain, and therefore lameness, when the weight of the body resting on the foot forces the sensible part against the nail.   This may also include wounds inflicted on the sole by broken glass, sharp flints, &c.

Symptoms.—Lameness, which, when the nail has entered the sensible part, quickly follows the injury, and suppuration succeeds, running under the sole.   When the quick is not actually penetrated, it is often some days before the lameness is visible.

Treatment.—Remove the shoes.   If one is wet, cut down on that sole until the injury is exposed.   If the lameness is slight, treat with chloride of zinc, one grain to the ounce of water, or apply a little tincture of myrrh, and if granulations or little lumps appear, apply a caustic to them, such as butyr of antimony.   If the lameness is great, and much injury is discovered on paring out the foot, put the foot in a linseed meal poultice, which should be renewed for some days before using any of the other applications.

PUMICED FEET.

Causes.—The result of founder, also battering the feet of horses reared on soft land on the stones of streets.

Symptoms.—Lameness; the sole bulges out, and comes in contact with the ground, and the animal treads on his heels.

Treatment.—A cure is not to be expected, but relief may be

E

given by putting a blister on the coronet, lubricating the feet constantly with equal parts of glycerine and tar ; and by the use of a bar shoe of the dish kind, with a leather sole.

QUITTOR.

Causes.—Confined matter from suppurating corns or prick of the sole, which rises upwards and finds its way out at the coronet ; or it may be caused by a tread, or any severe bruise to the coronet.

Symptoms.—Great lameness ; apply a poultice of linseed meal for some days, and afterwards inject a mixture of one part of carbolic acid and four parts of glycerine into the sinuses or "pipes." This is better than plugging up the sinuses with strong caustics, as some recommend, and if persisted in, along with repeated dressings of the ointment of carbolic acid, and rest, will prove of great benefit.

SANDCRACK.

Causes.—A false step ; treading long on very dry soil. Horses with dry, thin, and brittle hoofs more subject than others.

Symptoms.—A fracture or splitting of the crust, mostly on the inside, from one to two inches from the heels in the fore feet, and extending downwards from the coronet ; in the hind foot it is always found in front of the foot. Sandcrack in the hind foot occurs chiefly in heavy draught horses. The crack extends to the quick, causes it to bleed, and produces lameness.

Treatment.—Pare out the crack so as to convert it into a groove. Put the foot in a linseed meal poultice, after which draw a transverse line with a firing iron above and below the fissure, and a little gutta percha melted in and on it ; put on a strap round the foot, tolerably tight. Shoe with a bar shoe, and if possible give rest. If granulations sprout during any part of the course, dress frequently with chloride of zinc lotion, and then cut them off.

SPAVIN.

Cause.—Hard work : sickle or cow-hocked horses most subject to it.

Symptoms.—Deposit of bone upon the lower and inner side of the hock. The leg is imperfectly bent, and the toe is dragged along the ground. When the animal is turned, a certain degree of flinching will be detected. Exercise diminishes the symptoms, but on cooling down the stiffness returns.

Treatment.—Rest is the great essential. Apply cold applications to the part, and remove the shoes altogether. If the horse continues lame after a time under this simple treatment, recourse must be had to blisters, setons, and firing; but such active measures ought not to be adopted rashly, and only under proper advice. The repeated application of small quantities of the ointment of biniodide of mercury will promote absorption, but long rest and careful feeding are material points. Once a spavin is fully formed it cannot be removed.

SPLINTS.

Causes.—Putting young horses too soon to hard work ; blows. kicks, &c. If the bones of the leg are small in proportion to the carcass; if there is undue length between the knee and the fetlock; if the ligaments and tendons are small, or if the pasterns are over long or the legs crooked, and not placed well under the centre of gravity, there is every probability that splint will make its appearance.

Symptoms.—A deposit of bone, shown by swelling upon the inner and lower part of the knee of the fore-leg, or upon the shin bone of either limb. These swellings are painful when growing, shown when the splint is pressed, and cause lameness, the animal being unwilling to bend the knee as freely as before. In the early stage the growth may be checked by perfect rest and cold water bandages round the part affected. A three-quarter shoe is useful in lessening concussion. Apply the ointment of biniodide of mercury, well rubbed in, but in small quantities, not sufficient to cause serious inflammation. If the splint has become well formed, the best plan is to let it alone, if it does not cause lameness, but if it does the most successful treatment is the operation of periosteotomy, which must be done by a practised veterinary surgeon.

THRUSH.

Causes.—Moisture and filth, when in the hind feet, and sometimes in the fore feet by contraction and heat.

Symptoms.—An offensive discharge from the cleft of the frog, the surface of which becomes ragged, and the interior covered with a white powder. No permanent lameness, but if the animal treads on a rolling stone, he will fall as if shot.

Treatment.—Pare the unsound horn from the frog; tack on the shoe; apply a mixture of carbolic acid, one part, and glycerine, four parts, to the cleft, so as to clean it thoroughly, and leave a pledget of tow smeared with ointment of carbolic acid in the cleft; repeat while necessary; once the swelling is removed, a little liquor of lead will perfect the cure.

The following is the "stopping" employed at the Royal Veterinary College, and it will be found very beneficial in thrush and other diseases of the foot:—Take of common tar, 2 parts; soft soap, 1 part; ground linseed, a sufficient quantity to give tenacity to the whole. This is spread over the sole of the foot, about a quarter of an inch in thickness, and then a layer of tow is laid on it, and over all a leather sole is placed. Then tack on the shoe.

CAPPED HOCK.

Causes.—Kicking in the stable or in harness. Some horses also injure themselves in the act of lying down or getting up.

Symptoms.—A round swelling on the point of the hock, arising from a thin watery effusion under the skin at that point.

Treatment.—Foment and hand-rub; apply the ointment of

biniodide of mercury, lightly rubbed in, and repeated until signs
of an eruption appear, after which cease applying the ointment
until this effect has passed away. Capped hock produces no
serious effect, but it is unsightly and very difficult to get rid of.

Capped elbow and capped knee are of the same nature.

CURB.

Causes.—A sudden sprain in a race, an extraordinary leap, or
a severe gallop over heavy ground, &c. "Sickle hocks" are peculiarly liable to curb.

Symptoms.—An enlargement at the back of the hock, about three
or four inches under its projecting point. The swelling is not
great, but at first it is attended with lameness, which subsides
when the inflammation is reduced.

Treatment.—Rest, the period for which may vary from a fortnight to several weeks. If the horse is put too soon to work, the
ailment will return. Use cold applications as in ringbone (which
see), to reduce the inflammation; after which rub in the ointment of biniodide of mercury, or equal parts of iodine and cantharides ointment, or blister the part, and in some cases it may be
necessary to have recourse to firing.

RINGBONE.

Causes.—Concussion; a blow, tread, or other wound, or from
weakness or sprain of the ligaments of or about the fetlock.

Symptoms.—Ringbone is a bony deposit, either on the upper
or on the lower pastern bone, and more frequently found in the
hind than the fore fetlocks. Horses which have long pasterns,
or unduly short or upright pasterns, are most liable to it. The
lameness arising from ringbone is more perceptible on hard than
soft ground. There is a stiffness in the fetlock joint, and a consequent snatching up of the foot in action. There is also swelling and heat about the fetlock.

Treatment.—Rest, aided by cold applications, is the first thing to attend to. Cold water is the most handy application; but the following is very useful:—Take of common salt, nitre, and hydrochlorate of ammonia, equal parts; water sufficient to dissolve; wet linen bandages with it, and cover these with oiled silk. When the inflammation is reduced, rub in the ointment of biniodide of mercury, repeating until an eruption appears; then cease until this effect has passed off, and repeat again and again, with much friction, while the horse continues lame. If it shall be considered necessary to fire for this disease, use the budding iron, as the marks left by it are less perceptible than those made by the ordinary firing iron. Keep the toes of the shoes short, turn up the shoes at the toe, and use leather under the sole. A renewal of severe work will again set up inflammation in the parts, and cause a fresh deposit of bone.

SIDE BONES.

Causes.—Accidental blow, wound, or tread, causing inflammation of the lateral cartilages or wings of the bone of the foot, which, under the influence of inflammation, become absorbed and replaced by bone. Shoeing heavy draught horses with high calkins also gives rise to "side bones;" also hereditary predisposition.

Symptoms.—The horse is more lame on hard than on soft ground; but there is no special peculiarity about the lameness, except a certain degree of stiffness of action.

Treatment.—When the disease arises from inflammation caused by concussion or accidental wounds, rest and cold applications, as directed in ringbone, form the best treatment. Discontinue the use of high calkins, if their use has led to it, besides giving rest and using cold applications. A bar shoe will also be beneficial; but when once quite formed, side bones are incurable.

SPRAINS.

Causes.—Injuries caused by violent usage.

Symptoms.—In some cases inability to raise and bend the leg, and consequently a tendency to drag the toe on the ground. When standing still, however, the horse keeps the knee slightly bent, so as to relax the tendons. In other cases the leg is raised freely euough; but pain and flinchiug are shown when the foot comes to the grouud.

Treatment.—" The skill of the veterinary surgeon lies more," says Colouel Fitzwygram, "in detecting the exact seat of injury thau in the mode of treatment:" and he points out that many practitioners get over the difficulty " by blistering, firing, or putting setons all over the affected leg or legs, on the ground, we presume, that if the remedy be applied sufficiently extensively, it must somewhere or other hit upon the injured part." The great point in treating sprains is entire rest, which must be long-continued in severe cases. Apply cold water bandages or fomentation to the part. In the latter case the water should not be hotter than the hand can comfortably bear, aud the heat should be kept up to this point by adding occasionally small quantities of hot water. The fomentation must be continued until the inflammation is reduced. Let the man employed at the work be accommodated with a stool, and relieved from time to time. When the fomentation is discontinued, even for a short time, wrap up the legs in loose, woollen bandages. All corn must be withheld, and the horse prepared by cold bran mashes for physic, which may be given or not, according to circumstances. Put on a high-heeled shoe, to take off the weight, and in severe cases put the horse in slings. If he is fretful, sicken his stomach by the administration of a drachm of aloes daily for a few days. The period required to reduce the inflammation will vary from two or three days to ten days or a fortnight, according to circumstances. After the inflammation is reduced, apply a blister over and round the injured part; but if very severe, it may be necessary to fire the part or to put in setons. The horse must be put only very gradually to work.

The disagreeable habit of knuckling over behind generally arises from a sprain of the ligaments and tendons of the fetlock joint.

THOROUGH-PIN.

Causes.—Excessive labour.

Symptoms.—Thorough-pin is a further development of bog spavin, the swelling appearing equally on both sides of the back joint, so that the fluid contained in it—being an increased secretion of the synovia or joint oil—may, by moderate pressure, be forced from ohe side to the other.

Treatment.—See Bog Spavin. In all such cases rest is of absolute importance.

WIND GALLS.

Wind galls are enlargements in the neighbourhood of the fetlock joints, of a similar nature to bog-spavin and thorough-pin.

As a general rule, it is advisable not to apply treatment to enlargements of this kind. They may be eyesores, and sometimes even incapacitate the animal for fast work; but they seldom of themselves produce lameness.

BROKEN KNEES—OPEN JOINT.

Cause.—Fall on hard road.

Treatment.—If the skin or hair is merely rubbed off, it will probably be only necessary to sponge the part with lukewarm water; in doing so squeeze out the water above the knee, but do not dab the part itself. Rest for a day or two will be necessary; the application of a weak ointment of cantharides—one part to twenty parts or more of lard, applied with friction—will promote the growth of the hair.

If the knee is deeply cut or much contused, so that a slough takes place, Open Joint will be the result. This consists of an emission or running from the wound of the synovia or joint oil.

If the discharge is occasioned by the sheath of a tendon passing over the joint being cut through, the injury will not be serious; but if the joint is laid open, there is more danger.

In these cases the treatment must consist of perfect rest, laxative diet, and the administration of half a dose of aloes, or say two to three drachms of aloes, after the horse has been prepared. Take away all corn and hay. In order to secure perfect rest, put the animal in slings; but if that is inconvenient, reverse him in the stall, and secure his head by pillar reins, to prevent his lying down; and put a cradle round his neck, to keep him from biting the wound. Let the bed be made level, and if he is irritable and uneasy, give some slight doses of opium, say half an ounce of the tincture as a dose.

Wash the wound clean with lukewarm water, and apply linseed-meal poultices for the first twenty-four hours, but not longer. Professor Williams, of the Edinburgh Veterinary College, recommends that the next step should be the application of a blister, in order to produce such swelling of the neighbouring parts as may close up the opening and thus exclude the air from the joint. This treatment, it is said, is in general very successful.

The air may also be excluded by placing over the wound a fold of lint, kept constantly wet with cold water day and night, and secured lightly above and below, but not bandaged. In two or three days' time the cold water dressing may be changed to a slight astringent lotion, such as sugar of lead, half an ounce; vinegar, 2 ounces; water, 1 quart; mixed together; and afterwards a powder made of one ounce of sulphate of copper to a pound of flour may be lightly dusted over the wound. This will cause the formation of a scab, which will exclude the air. Do not remove the clot of synovia which will be found on the edges of the wound. As soon as the wound is closed, recovery will be rapid. If a case of open joint takes an unfavourable turn—that is, if the inflammation extends to the interior of the joint, and if matter begins to be discharged from the wound—treatment will be of little avail, and the more merciful course is to destroy the animal. The application of strong astringents and tight bandages are often causes of cases terminating unfavourably.

In the after treatment, give a little quiet led exercise twice a day, and repeated mild blisters, such as the ointment of biniodide of mercury, will prove beneficial in exciting the absorbents and thus removing any thickening of the parts. Dress any super-abundant granulations that may form round the wound with sulphate of copper.

Open joint in other parts must be treated in the same manner.

OVER-REACH AND TREAD.

Cause.—Fatigue, causing the coronet of the fore foot upon the outer side to be severely wounded by the inside of the hind shoe.

Symptoms.—A severe wound and large slough, which may be followed by " False Quarter." In tread it is the hind foot which is injured.

Treatment.—Rest, and bathe the part daily with chloride of zinc lotion; one grain to the ounce of water.

CHAPTER VIII.

SPECIAL DISEASES.

RHEUMATISM.

Causes.—Exposure to cold and wet, in which case it generally affects the loins and shoulders; insufficient diet; and bad stable management. It also frequently accompanies other diseases of a debilitating nature, such as influenza.

Symptoms.—Sudden and unaccountable stiffness in some parts of the body, usually the limbs; the affected part becoming inflamed and sore to the touch, accompanied by symptoms of fever, and short, quick breathing. The disorder flies about, or shifts from one part to another. When a part has been frequently attacked, a chronic swelling of that part generally takes place.

Treatment.—Apply friction to the part affected, and wrap it with flannel. Hot fomentations are also beneficial; but avoid a chill by drying the part well after fomentation, and rub it with ammonia liniment.* Give one ounce of bi-carbonate of potass, followed daily by a dose of half the amount with half an ounce of nitre, until relief is obtained. If these fail, give two drachms of iodide of potassium in addition. If the pain is great, give, instead of the above, one drachm of powdered opium and one drachm of aloes, with ginger and linseed meal, night and morning, for three days. Then discontinue the medicine for a few days, and afterwards repeat. If the pain is very great, add to

---

* The good effects of a *warm bath* may be attained in this and other diseases by the use of bags of warm, wet bran, laid over and across the loins, spine, &c., which, if covered over with rugs or blankets, retain the heat for a long time, and can be easily reheated by dipping them in very hot water, and wringing nearly dry before laying on the animal. When they are taken off permanently, the skin should be rubbed dry, and dry, warm rugs put on to prevent chill afterwards. This will be found very efficacious treatment in many diseases of horses, cattle, and also swine.

the above dose one drachm of extract of belladonna. Keep the bowels open; and for this end, give half a pint of castor oil, which repeat, if necessary. If lameness still remains after the inflammation has subsided, apply a blister of biniodide of mercury to the part, and repeat, if necessary. Care should be taken to have horses thoroughly dried and cleaned after returning from work or exercise; and in this way rheumatism and other complaints will be frequently prevented.

FARCY.

Causes.—The result of the presence of a poison in the blood, attended with a peculiar specific inflammation. The predisposing causes are bad ventilation, bad drainage, filthy stables, overcrowding; bad feeding coupled with over work and neglect in the stable management, such as not drying the animals on their return from work, &c. ; sometimes follows other diseases; contagion, infection, inoculation.

Symptoms.—Feverish symptoms; considerable swelling of one or more of the superficial absorbents, particularly of the hind legs, and in harness horses the neck or shoulder, ribs and lips. When farcy appears on the face, shoulders, lips, or sides, the buds are usually much smaller than when it appears on the hind legs. The formation of ulcers rapidly increases, and the body is sometimes covered with them. There is a smaller kind of farcy called " button farcy," which is usually the worst kind. The matter discharged from a farcy lump is either thin and of a dirty yellow colour, or it is like glue, or it may be bloody; but in all cases it is offensive. From the lumps little cords may be traced leading to other swellings. Appetite sometimes fails, at other times it is voracious, and a strong thirst exists.

Treatment.—There is little hope of success of any treatment in a confirmed case of farcy. If treatment is to be attempted, the animal should be kept strictly secluded from others, and attend carefully to good feeding and fresh air. Give biniodide of copper in doses of one or two drachms daily, or from 4 to 8 grains of cantharides, with two drachms of extract of gentian, or sulphate of copper in two drachm doses in a ball. If there is much swelling of the hind legs, give half an ounce to one ounce of nitre, combined with one to four drachms of sulphate of iron, and walking exercise. The buds or ulcers should be opened as soon as they are soft by the lancet, and dressed with the ointment of biniodide of mercury. The corded absorbents for some distance beyond the point to which the knots or ulcers extend should also be dressed with the same ointment.

Farcy generally runs into Glanders, which see. (See also Water Farcy.)

GLANDERS.

Causes.—Same as in farcy, of which glanders is usually the termination.

Symptoms.—First, a swelling in the gland under the jaw, on one side, and after a time a gluey discharge comes usually from one nostril. The gland becomes more swollen and painful, appears hard, and there is no disposition in the tumour to form matter and come to a head, or to disappear. The lining of the nose is brightened in colour. The discharge from the nostrils is at first of a watery nature, but it soon becomes mixed with a ropy slime. It afterwards becomes glairy, or like the white of an egg, but of a yellow appearance, and in the later stages it becomes of a more purulent character, like matter running from a wound. Whether the discharge is slight or copious, thick or thin, it is always like glue, and adhesive, clinging about the hair round the nostrils, and partly clogging them, so as to produce difficulty in breathing. The last symptom, which leaves no doubt as to the nature of the disease, is the formation of ulcers in the nostrils, after which the discharge is often tinged with blood, and is offensive. These ulcers rapidly increase, and extend along the throat and windpipe, and tubercles form in the lungs, which end in becoming abscesses, and these, when they burst, cause death by suffocation. In the later stages of the disease, the animal falls away in condition ; the hair of the mane and tail is easily pulled out ; there is generally a large development of farcy buds in various parts of the body, and the animal dies a mass of disease.

Treatment.—Incurable, although in the very earliest stages the treatment given for farcy may succeed in arresting its progress. It is, however, so dangerous, even to the attendants, that a glandered horse should be destroyed at once, and either buried deep, without being skinned, and covered with hot lime, or put into the vat of the manufacturers of superphosphate.

In both glanders and farcy it is desirable to burn any articles of clothing, brushes, brooms, &c., to which diseased matter may adhere. Powerful disinfectants, such as carbolic acid or chloride of zinc, should be applied liberally and repeatedly to all wood-work, floors, &c. After these have been thoroughly scraped, all woodwork should be re-painted. Leave every door and window open for a considerable time, during which no other animal of the kind should be allowed to enter. In fact, it is impossible to be too careful when either farcy or glanders appears, or has existed in a stable.

Other diseases are sometimes apt to be mistaken for glanders, such as catarrh, pneumonia, nasal gleet, strangles in its early stage, the effects of a rotten tooth, and even ordinary sores of the nostrils. In all suspicious cases the full development of the disease should be waited for, the patient being, of course, kept strictly separate from all others ; and it may be advisable to apply a test by inoculating an old ass, and if the disease is actually glanders, it will, when germinated by inoculation, be fully developed in from six to fifteen days. If the wound heals without producing the effect, it may be concluded that the disease is of some other nature.

WATER FARCY, WEED, SHOT OF GREASE.

These names are all applied to a disease of the absorbents ;

that is, the minute vessels which return the watery part of the blood to the heart. These, when overcharged with watery fluid, or when inflamed, become diseased. Draught horses are more subject to disease of this kind than any other, and those which are known as " gummy-legged" animals are much predisposed to it. It is important, however, to keep in view that the occurrence of this form of disease indicates culpable neglect in stable management, which, if not amended, will lead to true farcy and to glanders.

Cause.—Debility in the system, which produces certain effects when hard work and long hours are followed by a state of continued rest, such as from Saturday night till Monday morning, in the vitiated atmosphere of a closely shut-up stable.

Symptoms.—One of the legs, generally a hind one, becomes suddenly swollen, the swelling being of a dropsical character; there is a painful degree of lameness; pain inside of the thigh, which is increased upon pressure; the appetite is gone, but the presence of fever is indicated by an insatiable thirst. If the case is neglected or improperly treated, the leg becomes permanently swollen.

Treatment.—In slight cases a little gentle exercise will often remove the symptoms, for which reason horses manifesting any tendency to it should not be allowed to remain long in the stable without exercise, and half an hour's led exercise on Sunday will often keep an animal of that kind right. When this is not sufficient, prepare the horse for physic by depriving him for at least 36 hours of all food, except cold bran mashes; then give a purgative ball, say powdered aloes, 3 drachms; powdered gentian, 3 drachms; treacle sufficient; mix, and when this is worked off, follow with diuretics and tonics. Diuretic balls are generally made of two drachms each of resin and nitre, with one drachm of Venice turpentine, mixed in a mass with soft soap and linseed flour. The best tonics are good, fresh, cool air, well regulated exercise, good feeding, and good grooming. Both mineral and vegetable tonics are dangerous, if the use of them is pushed too far, and their action requires to be very carefully watched. Of mineral tonics the ordinary dose may consist of—Sulphate of iron 1 to 2 drachms, with 2 to 4 drachms of ginger; or sulphate of copper, ½ to 1 drachm, with 2 to 4 drachms of tincture of gentian. Of vegetable tonics the ordinary doses are—Quinine, ½ to 1 drachm, dissolved in a few drops of sulphuric acid and a pint of water; or oak bark, 2 to 3 drachms, made into a ball with treacle and bran; or tincture of gentian, 1 to 2 ounces, in a pint of water.

In water farcy or weed, if the lameness continues, it may be necessary to make a few punctures in the skin. Apply cold water freely to the limb every morning, but after doing so let the part be thoroughly dried by hand-rubbing.

PARTURITION.

Mares usually foal easily, quickly, and seldom require assistance; but although they continue to work until the immediate symptoms of foaling appear, they should, as the time draws near, be put into a roomy loose-box, and watched during the night, in case of accidents. The proper position for the fœtus or unborn foal to lie in the womb is with the fore feet extended backwards, and the mouth just above them; so that, in escaping from the womb, the feet should make their exit first, and act like a wedge in dilating the mouth of the womb and other parts. It is easy, therefore, to conceive how great must be the inconvenience arising from a wrong presentation. When the hind parts come first, it is evident the larger portion of the body will have to escape first, instead of the smaller; and this is greatly increased if the hind feet, instead of being extended backwards, are advanced forwards under the body. Another inconvenient posture is, when the fore feet are extended towards the mouth of the womb; but the head turned backwards. When these cases occur insert the hand and push back the foal, and bring forward the parts that ought to come first. It is sometimes necessary to pass a cord round the legs, by pulling which the efforts of the mare are assisted. Much patience, and also cleverness, are sometimes required in difficult cases of foaling, and in many instances the foal has been lost from impatience and over eagerness. The same remark holds good in the case of cows, &c.

When the foal is dead, labour is more difficult, and it is sometimes necessary to remove the foal piecemeal. If the womb does not contract, and there are no labour pains, give a dose of two drachms of ergot of rye in a little gruel, and this will have the effect of stimulating the action of the womb.

In all cases let the mare's diet after foaling be of a laxative nature, and for the first day or two her drink should consist of a little oatmeal, mixed with lukewarm water. Let her go to the grass after a few days; but if the pasture is not sufficiently advanced, or the weather unfavourable, let her have boiled turnips at night, or carrots, along with some oats, and warm bran mashes during the day.

Inflammation of the womb sometimes occurs, but not often. When it does occur, bleed, give laxative medicines, say 2 pints of linseed oil to which 5 drops of croton oil have been added. Give also sedative medicine, such as tincture of aconite, 10 to 20 drops, until the pulse is relieved; also injections.

MELANOSIS.

Cause.—Unknown.

Symptoms.—This disease is confined to gray horses which have become white. It consists of a number of tumours on various parts of the body, but frequently attached to the tail. These tumours when cut into are found to contain a black fluid or substance.

Treatment.—When this disease is confined to the tail, the tumours may be cut out, but all treatment is useless when they appear elsewhere. In dressing the animal let the brush only be used. Mayhew recommends the body to be anointed twice a week with animal glycerine, one part; and rose water, two parts.

F

SPRAIN OF THE LOINS.

Cause.—Violent over-exertion, causing sprain of the ligaments of the back and loins.

Symptoms.—The horse rolls in his hind quarters, and has no proper control over the muscles.   When the injury is slight the animal may move pretty well in a straight direction ; but the injury will be detected when he is backed or turned sharply round.

Treatment.—Long rest, and repeated blisters.   Adhesive plaisters, composed of Burgundy pitch, softened with a little oil of turpentine, have been found useful.   In some cases recovery is so uncertain that it is better to destroy the animal.

FISTULOUS WITHERS.

Causes.—Pressure from the saddle, or collar, or by injuries.

Symptoms.—A small, round swelling appears on the spot affected, and if this is neglected, the place enlarges, and numerous holes burst out, which are the mouths of so many fistulous pipes.

Treatment.—If the swelling is slight, give the horse rest from work for a few days, and dress the part with a little salt and water, and if the swelling enlarges, apply fomentations.   Should these fail, matter will form under the skin, which must get vent. Lay the part open with the knife, and if the injury is deep seated insert a seton through the sinus or pipe.   Dress the wound with a solution of sulphate of zinc.   A mixture of one part of carbolic acid and six parts of glycerine injected into the wounds will be useful ; and the ointment of carbolic acid may be smeared on a pledget of tow and inserted in it ; remove, and put in a fresh pledget at least twice a day.   (See " Poll Evil.")

POLL EVIL.

Causes.—Pressure of the head collar ; an accidental blow, such as a horse may give his head in passing through a low doorway ; a blow on the head with the butt-end of a whip.

Symptoms.—A swelling on the top of the head just behind the ears. If this is neglected, abscesses will form. When feeding, the horse hangs back from the manger, keeps the nose protruded, and the head as motionless as possible.

Treatment.—See "Fistulous Withers," which poll evil resembles in its nature, and the general treatment required. In either case the application in the early stage of a light blister, such as tincture of cantharides or the ointment of biniodide of mercury, if applied in time, may prevent the spread of the disease.

CRIB-BITING.

Causes.—Some acidity or chronic irritability of the stomach in the first instance, and afterwards a bad habit.

Symptoms.—The animal catches at the manger with his teeth, arches his neck, and sucks in a quantity of air with a peculiar noise. After a time the abdomen is evidently enlarged.

Treatment.—See that the ventilation of the stable is pure. Give a change of and variety in the food. Place a lump of rock salt in the manger, and a piece of chalk. Damp the corn and sprinkle magnesia upon it, and also mingle a little good oak bark with each feed of corn. If crib-biting becomes a habit, put on a strap round the neck, which must be kept sufficiently tight to prevent the horse from swallowing the air. Another remedy is a muzzle made for the purpose. It consists of an iron rack so wide as to allow the horse to seize his food and yet so narrow as not to permit the passage of the teeth. Coating the manger, &c., with oil of tar has been recommended as a preventive.

Wind-sucking is similar to the above, and requires the same treatment.

INFLAMMATION OF THE JUGULAR VEIN.

Cause.—Bleeding in the neck with a blunt or foul fleam or lancet, or by repeated wounds of the part in attempting the operation, and allowing hairs or other foreign bodies to interfere with the proper adjustment of the edges of the wound.

Symptoms.—The first symptom is a separation of the cut edges of the integuments, which become red and somewhat inverted. Suppuration soon follows, and the surrounding skin appears swollen, tight, and hard, and the vein itself above the orifice feels like a hard cord. After this the swelling of the neck increases, accompanied with extreme tenderness ; and there is constitutional irritation with a tendency to inflammatory fever. The head and neck become swollen on one side, and death may be the consequence.

Treatment.—In the first stage try to relieve by cold lotions or hot fomentations. If these fail, and as soon as the disease begins to spread in the vein, touch the spot with lunar caustic, or even the hot iron, simply to sear the lips of the wound, and apply a blister over it, which may require to be repeated. Purgatives in full doses must be administered, and the neck kept steady and upright.

Sometimes after bleeding, a globular swelling as large as an egg arises immediately around the newly made incision. Apply gentle pressure with a sponge and bandage, kept cool with cold lotions. This affection, although alarming enough in appearance, is in reality very trifling.

PART II.

DISEASES OF CATTLE.

CHAPTER I.

THE HEAD.

APOPLEXY

Is a determination of blood to the head, and may be attributed to a redundancy of flesh and fat. It is frequently rapidly fatal.

Treatment.—Copious bleeding, followed by active purgatives, is the only remedy; but blisters to the back of the head may also be tried. (See "Splenic Apoplexy.")

INFLAMMATION OF THE BRAIN.

Cause.—This disease is frequently brought on by over-driving, and bad treatment, combined with excitement.

Symptoms.—Great excitement and extreme violence is manifested, though this state is sometimes preceded by one of heaviness and oppression. There is reason to believe that this disease is frequently mistaken for rabies, or madness produced by the bite of a mad dog.

Treatment.—Copious blood-letting and physic are the only remedies; but it is often impossible to have the animal bled, owing to the furious state in which it is while labouring under the disease; but physicing may be accomplished by taking advantage of the thirst of the animal, and supplying Epsom salts in the water in large quantities. In an advanced stage of the disease the blood will be so congested in the smaller veins as to render the flesh unfit for human food. (See "Stomach Staggers.")

BLAIN.

This name is used in many parts with reference to murrain, or foot-and-mouth disease; but it is also employed as a common designation for the disease known to veterinarians as "Gloss-anthrax"—a contagious disease, the seat of which is the tongue.

Causes.—The direct causes are not well ascertained; but it is known that it is distinguished by a blood poison, and that it is developed under the combined influences of heat, moisture, and putrefactive emanations from the soil.

Symptoms.—The animal refuses to eat; saliva flows freely from the mouth; the lips, cheeks, and neck swell; breathing becomes laboured; the discharge from the mouth becomes offensive, tinged with blood, and of a greenish colour. The tongue, which is swollen and raised, is covered with small blisters, which extend also to the lips and gums; these enlarge rapidly, and ultimately burst. The animal dies in from twenty-four hours to three days.

Treatment.—Open the blisters, and wash the mouth freely and repeatedly with vinegar and water; and in severe cases, touch the blisters with nitrate of silver. Strong purgative medicine must also be given; and if the animal begins to recover, feed her on mashes, and a little malt daily.

CHAPTER II.

THE THROAT AND RESPIRATORY ORGANS.

~~~~~~~~~~

CHOKING.

Since machines for cutting up roots into small slices—" finger pieces"—have come into use, and, better still, since root pulpers have been employed in preparing the food of cattle, choking has become of comparatively rare occurrence. In fact, the use of such machines prevents it, as it only takes place when the roots are given whole, or when the slices into which they are cut are too large.

Symptoms.—Evident distress ; difficulty in breathing ; an attempt to vomit ; a discharge of frothy saliva from the rumen or paunch, and swelling as in hoove.

Treatment.—First, pour down the throat a little oil, and then introduce the probang, which should be greased. When it has reached the obstructing body, use firm and moderate pressure, and alternately raise and depress the head. If it does not easily yield, do not be impatient, and when pressure is again applied, the obstruction will quickly be removed. Feed the animal for some days on mashes, avoiding solid food. If a probang is not at hand, the handle of a cart whip may be used, being first greased ; but great caution must be exercised in so doing. Where prevention is so easily attained by means of the machines referred to above, choking need not occur.

TUMOURS, OR "CLYERS," IN THE THROAT.

Professor Dick gave an account (see " Veterinary Papers") of his treatment of certain *encysted* tumours situated for the most part in the throat at the superior angle of the lower jaw, but occasionally appearing in the back, flank, and tongue, from which we´take the following particulars :—

Causes.—Professor Dick attributed them to want of proper shelter, either natural or artificial, combined with constitutional weakness, more especially as evinced in the case of cattle which have been closely bred.

Symptoms.—The tumours are sometimes superficially attached to the cellular membrane of the lower jaw, and when such is the case are easily removed by excision ; but for the most part they are of a malignant nature, forming an abscess, which is generally of an indolent character. Young heifers are most susceptible of the disease.

Treatment.—Confine the animal to a comfortable house, and increase the strength of the system by nutritious diet, combined with tonic medicines, such as the sulphate of iron in doses of 2 drachms to a middle-sized two or three-year-old, once a day, in a little gruel, or 5 grains of iodine and gruel, night and morning. The tumour should be laid open if matter is felt ; but if not, the ointment of iodine, 2 drachms to 4 ounces of lard, should be well rubbed into the tumour every day until the absorption of the tumour takes place. Should this fail, open the tumour with the lancet, and put a little tow into the wound, twice a day, smeared with blistering ointment, or dipped in a strong solution of sulphate of copper.

BRONCHITIS.

This disease is often the consequence of neglected catarrh (see "Catarrh"), but more dangerous, as it affects the internal surface of the lungs.

Symptoms.—Similar to, but more severe, than those of catarrh, and there is a greater soreness in the act of coughing.

Treatment.—Bleeding in the early stage ; a seton inserted in the brisket ; mustard poultices over the neck and chest ; or external stimulating liniment given under catarrh ; mild, purgative medicine ; doses of tartar emetic, $\frac{1}{2}$ drachm to 1 drachm, mixed with 2 to 4 drachms of nitre ; warm mashes ; comfort. (See " Hoose".)

CATARRH—"COMMON COLD."

Causes.—This disease is peculiarly prevalent in exposed situations, and occurs chiefly in the spring of the year, when the wind is easterly, particularly if the weather is wet as well as cold. It also occurs in wet weather in the autumn, and among young beasts kept in the fields during winter without proper shelter from cold and wet.

Symptoms.—A cough, dry muzzle, horn hot, a little heaving at the flank, constipation, and high coloured urine.

Treatment.—In slight cases, housing, with a dry bed, and a few warm mashes with a little nitre in them, will put all to rights ; but when the case is severe, then, in addition to housing, bleed moderately, and give a dose of Epsom salts—$\frac{1}{2}$ lb. to 1 lb.—with $\frac{1}{2}$ ounce of powdered ginger in it, and some treacle, and rub well the following stimulating liniment upon the skin of the throat :—

Powdered cantharides	... ...	1 ounce.
Olive oil	... ...	6 ,,
Oil of turpentine ..	... ...	2 ,,

Mix. A seton inserted into the brisket is also very serviceable.

Catarrh sometimes assumes an epidemic form, in which it causes great debility. In addition to foregoing treatment, give doses of gentian and ginger mixed in equal parts, from one to four drachms of each, according to the age of the animal. It is great folly to overlook catarrh in its earlier stages.

CONSUMPTION.

Consumption prevails chiefly among aged cows, and is frequently found in the crowded, ill-ventilated cow-houses of town dairymen.

Symptoms.—Oppressive cough ; dulness ; dainty appetite ;

emaciation ; hidebound; staring coat. Cows with this disease stand obstinately with an arched back and dejected look. There is a discharge from the eyes and nose, with diarrhœa. The milk is usually yielded abundantly, but it is blue and watery.

Treatment.—The treatment lies chiefly in giving rich, but easily digested food, such as linseed cake, palm-nut meal, and a little malt mixed with each feed. Careful nursing in every respect will be useful ; but the disease is incurable.

HOOSE.

Youatt classes Hoose under the same head as Catarrh, but Mr. Spooner, in "Morton's Cyclopedia," describes it as a distinct disease, to which young cattle, and particularly calves, are subject.

Symptoms and Causes.—A dry, husky cough; increased respiration ; great weakness and indisposition to move. These symptoms, says Mr. Spooner, are caused by the presence of small, white worms in the windpipe and bronchial tubes, which keep up constant irritation. The worms are produced by eggs supposed to be taken with the water, and it has been caused by keeping the calves on very bare pasture during the summer months. This disease has sometimes been termed Bronchitis (see "Bronchitis"), and Professor Gamgee designates it " Pthisis pulmonalis verminales."

Treatment.—Give half a pint of lime water every morning, and a tablespoonful of salt to each calf every evening, continuing these doses for four or five days. The disease has also been treated successfully by giving an ounce of oil of turpentine in four ounces of linseed oil, and repeating the dose once a week several times.

PLEURO-PNEUMONIA.

This disease is of foreign origin, and there is every reason to believe that a single case of it has never spontaneously originated in the British isles. There is a kind of disease termed also pleuro-pneumonia, which is non-infectious, and to which man and many, if not all, the domesticated animals are liable. This disease is, however, not contagious, nor does it, as a general rule, run a fatal course. Both forms of lung disease are, so far as external symptoms are concerned, somewhat similar; but even before death an experienced observer is able to discriminate between them.

Causes.—Contagion.

Symptoms.—As in all contagious affections, a period of incubation takes place. This is simply the time, as already alluded to, that elapses between known contact with diseased animals and appearance of the signs which mark the disease. It has been stated by some that the symptoms may appear in a fortnight. They, however, more generally occur in about six weeks or forty days. In some instances a further delay takes place, as the transmission to Australia, occupying a voyage of three months, would appear to bear out.

During the period of incubation the animal is said to be *infected.* The condition frequently improves, and the yield of milk likewise increases—facts which may not be inaptly attributed to the primary effects of the poison upon the nervous system. The first symptoms are usually obscure and unnoticed, except by professional or practised eyes. These are slightly disturbed states of the respiratory and circulatory organs, accompanied by slight shivering fits and elevation of the temperature of the body, as indicated by Cassella's registering thermometer. Occasionally a cough is heard, which is not, however, sufficient to cause alarm, unless there be in addition loss of appetite, hot mouth, dry muzzle, and decrease of milk.

The milder signs continue for several days, when the bowels become costive, cough more troublesome and persistent, with staring coat, and intervening shivers. The pulse is tolerably full and rapid, numbering 80 or 100 beats, but shortly becomes smaller and not so distinct. The respiration is accelerated, and to facilitate this, as the disease advances, the nose is extended and neck brought into a straight line. A moan or grunt accompanies each inspiration, which, if it comes on at an early period, may be accepted as corroborative of a critical state. The nostrils are dilated, and the sides drawn rapidly backwards and forwards by the action of the air. A thin, mucous discharge sometimes flows from the nose and the eyes, and the latter, becoming bloodshot, sometimes discharge from the first a muco-purulent secretion. The horns, ears, and extremities become cold, urine deficient and highly coloured, mouth clammy, and skin tightly bound down, feeling dry and harsh to the fingers. If the spaces between the ribs or the region over the spine are pressed, pain is evinced, and not uncommonly occasions a low, deep moan. When the ear is placed in front of the windpipe a loud rushing noise is heard in the passage, by the transit of air through the tubes, the lining membranes of which are thickened, inflamed, acutely sensitive, and non-secreting. Louder sounds are also to be detected in the anterior parts of the lung, which are perfect; while respiration in the latter is silent, as a result of consolidation. As the lungs move by the process of respiration, assisted by the abdominal and other muscles, the surfaces, coming in contact with the sides of the chest, give rise to a sound not unlike the twisting of leather. This is often very general, and depends upon the extension of inflammation throughout the serous membrane.

In some cases only one lung is affected, when the respiratory murmur in the sound organ is much louder, in consequence of the increased amount of labour exacted from it. But when both lungs are affected, peculiar changes and complications set in as the disease advances. Portions may become gangrenous, or cavities are formed within the substance as a result of the process of suppuration, when the sounds called *ralés* are heard, which materially differ in accordance with the size of the cavity, its opening, and nature of contents.

In some favourable cases the lungs gradually resume their

action, on the removal, by absorption, of the tissue thrown out by the process of inflammation. Progressive recovery takes place, the pulse becomes fuller, respiration slower and more regular, and the healthy murmur tolerably well established. At other times complete destruction rapidly ensues on the appearance of consolidation. Water is thrown out from the surfaces, and accumulates within the chest, and the animal dies, as it were, by a process of internal drowning.

There are other cases which indicate a tendency to protracted duration. Little progress either towards death or recovery takes place. Discharges of a fœtid nature issue from the nose and eyes, appetite is capricious, and the animal becomes poorer every day. A troublesome sore cough continues, and not uncommonly portions of the disorganised lungs are expelled, with putrid pus. The temperature of the body is extremely low, and great prostration present. The animal is in a state of hectic or consumption, and speedily a fœtid diarrhœa, with typhus symptoms, usher in death.

After death the lungs are found much increased in weight, and more or less solidified. Mr. Armatage ("Transactions Highland and Agricultural Society, 1870") states that he found the lungs in one case to weigh 75 lbs. We ourselves have found the weight of lungs to exceed 100 lbs., or about ten times their normal weight. The increased weight is due to an albuminous matter which deposits itself in the cells or little cavities in the lungs, and becomes hardened therein. When a piece of hepatized (liver-like) lung is cut across, the sections present a mottled appearance, and not unlike that of sienna or coloured marble. Sometimes large quantities of purulent matter are contained in the lungs. We have often seen more than a gallon of this horrible stuff expressed from the lungs of a single animal. One of the characteristic systems of the disease is the adhesion of the pleura or bag surrounding the lungs to the ribs. As a rule, the more extensive the disease, the greater is the amount of adhesion.

Treatment.—With respect to the treatment of lung disease,

pleuro-pneumonia of the infectious kind is really a hopeless malady. The disorganization of lung tissue is so extensive that it is most unlikely an animal could live with an enormous proportion of its breathing organs solidified and rendered wholly unable to perform the respiratory function. In the early stage of the disease, when the lungs are merely congested or inflamed, but not hepatized, something might, indeed, be done; but, as a general rule, solidification of the lungs commences immediately after the animal is noticed to be ill; and then the disease assumes such serious proportions that it appears to be beyond the control of any drug, however potent.

Mr. John Gamgee, while saying that it is impossible to eradicate the disease from the system, states that "mustard poultices or more active blisters have been attended with good success. Stimulants have proved of the greatest service;" and the late Professor Lessona, of Turin, strongly recommended, from the very outset of the disease, the administration of strong doses of quinine. Maffei, of Ferrara, states having obtained great benefit from the employment even of ferruginous tonics and manganese in the very acute stages of the malady, supported by alcoholic stimulants. Recently the advantages obtained from the use of sulphate of iron —dose, 2 to 3 drachms—both as preventive and curative, have been extolled in France.

Preventive Measures.—Strict supervision of the cattle traffic; avoidance of contagion; the slaughter of infected beasts; the disinfection of stalls vacated by slaughtering; the closing of infected places to all passage of cattle; especial attention to the removal of the dung and remains of infected beasts. (See "Inoculation," *Index.*)

PLEURISY.

This disease is comparatively rare, and when it does occur is generally complicated with pneumonia or inflammation of the lungs. In the early stage the treatment is very similar to that

practised in cases of pneumonia (see " Pneumonia.")   When the
disease goes so far that there is an accumulation of water in the
chest, it will be necessary to make an opening with a small trochar,
but this is an operation which should be left to the qualified veterinary practitioners.

✕ PNEUMONIA. ✕

Causes.—Sudden change of situation, as from a dry to a damp
locality; over exertion when the animal is in a state of plethora.
Professor Dick gives an account (" Veterinary Papers") of certain
cases of pneumonia in cattle which were caused by feeding animals,
kept in open yards exposed to the east, on hot food three times a
day.   That it arose from the cause stated by Professor Dick was
evident from the fact that none of the cattle on the same farm,
" in adjoining yards and sheds, although 110 in number, not
being fed in a similar manner, nor being exposed to the cold
winds, exhibited symptoms of the same kind of disease."

Symptoms.—Quick and laborious breathing; mouth hot, but the
horns, ears, and feet are often excessively cold; and a frequent
sore cough.   The animal is generally found lying down, being
unable or unwilling to rise.

Treatment.—Extensive bleeding, purgation in the first instance
by means of Epsom salts—½ lb. to 1 lb.—and kept up by doses
of sulphur—six to twelve ounces.   A strong blister should be
applied opposite the lungs, but if this is ineffectual we have found
the following plan successful.   Steep a horse rug in hot water,
lay it on the side of the animal, put a dry rug over it, then
another steeped in hot water, and again a dry rug over all.
Setons inserted in the dewlap will also be useful along with the
other treatment.

CHAPTER III.

THE STOMACH, LIVER, SPLEEN, BOWELS, KIDNEYS, BLOOD.

INFLAMMATION OF THE PAUNCH—POISONING.

This is sometimes caused by eating poisonous plants; amongst which are enumerated the hemlock, water dropwort, henbane, wild parsnip, the yew, especially when withered and dry, and even the common crowfoot and the wild poppy. Cattle are, no doubt, endowed with a fine sense of smell, which generally enables them to reject food which is unwholesome, but occasionally, and especially when very hungry, they will eat ravenously of plants that are unfit for use. For instance, there have been cases where cattle have been poisoned in consequence of eating some of the plants named on being turned into a field where these grew luxuriantly, after a long journey by rail during which they had got neither food nor water. Ergot of rye, or the same poisonous growth when it appears in rye-grass, &c., produces abortion in in-calf cows when only moderately eaten, and death when it has been largely devoured. In the case of poisoning from the above sources, emptying the paunch should be attempted, also bleeding, and acidulated draughts.

But there are other forms of poisoning besides that caused by certain plants. Where dung collected in cities is used for top-dressing pastures, &c., cattle are frequently poisoned from swallowing bits of lead which had been mixed with the manure. Again, some cases of poisoning occurred in Scotland from cattle grazing in a field where there was a sand-pit nto which a quantity of rubbish had been thrown. The rubbisn consisted chiefly of green paper which had been stripped off the walls of an adjoining house, preparatory to fresh papering the same. The old paper was chewed by some of the cattle grazing in the field where it was thrown, and the arsenic, with which it was largely impregnated, killed them. The use of dressings containing arsenic or corrosive sublimate, for mange and other skin diseases, has also been the means of poisoning cattle. The best antidote for poisoning by corrosive sublimate is the white of several eggs beaten up and given in gruel. For arsenic poisoning, give lime water, but usually the effects are so rapid that there is not time for treatment.

STOMACH STAGGERS.

Professor Dick records ("Veterinary Papers") certain cases which came under his notice of inflammation of the abomasum, or fourth stomach, which he attributed to a sudden change in the condition of the food, as, for instance, turning cattle out from the straw-yard in which they had been wintered, upon rank grass; also a rapid growth of grass consequent on rain after a long tract of dry weather, combined with scarcity of drinking water, or its being of bad quality. The animals so affected had a small pulse, varying from 80 to 100; a strong inclination to press forward with the head against anything that came in their way, but especially in a corner; and at times they were in a state of high delirium, scrambling upon the walls with their fore feet; their respiration rather hurried, appetite gone, bowels inactive.

The *treatment* followed by Professor Dick was repeated bleedings, even to faintness, and large doses of purgative medicine given in large quantities of gruel, and cold water applied frequently to the head. The animals so treated recovered.

SWELLING OF THE STOMACH FROM FOOD.

This is different from hoove (see "Hoove"), although resembling it in some respects. The abdomen is distended, but it does not sound like a drum when struck, as in the case of hoove. It

G

usually occurs in stall-fed beasts, especially where fed largely on oats or oatmeal.

Treatment.—In mild cases a dose of castor oil or linseed oil, one to two pints, will relieve the animal, and stimulant remedies, such as oil of turpentine, &c., will occasionally prove effective, but the chief reliance must be placed on drenching the animal plentifully with tepid water, and the injection of the same into the stomach by means of the stomach pump. In some cases it may be necessary to make an opening through the flank into the paunch, sufficiently large to introduce the hand, and thus remove the contents. The wound must then be stitched up, but this operation should only be attempted by a practised hand, as without great care part of the food may escape into the abdomen, where it will produce great irritation and terminate fatally.

Some cattle appear to have naturally a tendency to indigestion, or debility of stomach, which causes them to swell frequently, and in such cases the greatest attention must be paid to diet, and tonics should be occasionally administered, such as gentian and ginger, of each one to four drachms, mixed together.

FARDELBOUND.

This is a disease of the maniplus, or manifolds, and is in some respects akin to red water.

Symptoms.—Pulse high, but respiration not much quickened; muzzle dry; mouth hot; the tongue hangs out, and seems enlarged; the eye is protruded and weeping; the head extended; the animal unwilling to move; the gait staggering; urine, which is sometimes red, or even black, is voided with difficulty, and there is obstinate costiveness; any hardened excrement which is voided has a bad smell. The disease is sometimes rapid, but at other times the animal may linger for two or even three weeks.

Treatment.—Pour down slowly a dose of Epsom salts, 1 lb., and 1 lb. treacle, in warm gruel; and either pour down slowly, or by means of the stomach pump keep up an almost constant current of warm water, having salt and treacle dissolved in it, but avoid all heating or stimulating medicine. Even the portion of ginger mixed with the dose of Epsom salts must be small. Give then warm gruel without stint, but no other food, except very thin mashes. Repeat salts, &c. Injections do very little good. It is a difficult disease to overcome.

HOOVE, OR BLOWN.

Causes.—This arises from an accumulation of gas in the paunch, caused by active fermentation of food in the stomach. It usually follows feeding on damp or luxuriant clover. (See "Swelling of the Stomach from Food.")

Symptoms.—Uneasiness, pain, and a swelling on the left side of the belly, which, when struck, sounds like a drum.

Treatment.—Give at once the following draught :—

Powdered ginger	... ...	3 drachms.
Hartshorn	... ...	1 ounce.
Water	... ...	1 pint.

In the event of these medicines not being at hand, give gin, whiskey, brandy, or turpentine, in one or two ounce doses; or some lime water; or, what is better, two drachms of chloride of lime dissolved in a quart of water. Pass down the hollow, flexible probang into the stomach, in order to allow the gas to escape, and follow up the medicines as directed above with a dose of Epsom salts, treacle, and ginger. It is sometimes necessary to make an incision on the left side, between the last rib and the hip bone, to allow the gas to escape. This is done by means of a trocar, which, when withdrawn, leaves a tube in the hole, which is retained until the gas has all escaped. A simple process is followed in some parts when an animal is "blown." A pailful of water is dashed on the back or loins, which causes the wind to be dispelled by belching, and if one pailful does not succeed, in five minutes another is dashed over her. We know this treatment has been successful in several instances.

Professor Dick recommends the following mixture of the oils of turpentine and linseed as nearly a specific in hoove :— Linseed oil, raw, 1 lb. ; oil of turpentine, from 2 to 3 ounces ; laudanum, from 1 to 2 ounces, for one dose ; or, hartshorn, from half an ounce to 1 ounce, in 2 pints of tepid water. In cases of pressing urgency, from 1 to 2 ounces of tar may be added to half a pint of whiskey, and given diluted, with great prospect of advantage. Remedies cannot be too soon applied in cases of hoove.

COLIC.

Colic in cattle is generally of the flatulent kind.

Symptoms.—Evident suffering ; the animal moans, strikes at the belly with the hind feet ; there is a swelling on the right side of the belly ; an occasional discharge from the mouth and fundament ; constant restlessness, and fever.

Treatment.—Give two drachms of the chloride of lime, dissolved in a quart of warm water, to which add two drachms of powdered ginger and twenty drops essence of peppermint. Walk the beast quietly about, without worrying it with dogs, as is sometimes done. Should the animal not be relieved in a quarter of an hour, give a warm purgative drink, and also injections of warm water or thin gruel ; rub the belly and flanks with spirit of turpentine. The food for some time after an attack should consist of warm mashes, warm gruel, and good old hay.

SPASMODIC COLIC is distinguished from the flatulent form by the smaller quantity of gas expelled, the comparative absence of enlargement of the belly, the spasms relaxing for a little time and then returning with increased violence, and the freedom with which the animal moves during the remissions.

Treatment.—Bleed in the first instance, and then follow up with purgatives and injections.

CONSTIPATION

Sometimes exists as a disease itself, at other times it is a concomitant of other diseases.

Treatment.—Aperient medicines with stomachics, such as Epsom salts, treacle, and ginger, castor or linseed oil. When obstinate, give from 4 to 10 grains of croton seed, along with the Epsom salts or oil. (See "Fardelbound.")

LOSS OF CUD.

Although loss of cud, or the cessation of rumination, is usually considered a disease, it is more a symptom of disease than a disease of itself. It does sometimes occur, however, without the appearance of any decided disease, and when this is the case, give the full grown animal a pound of Epsom salts with an ounce of carraway powder and half an ounce of ginger mixed with it in a quart of warm ale or gruel. This will usually put matters right; but should there be any fever, omit, or greatly reduce, the quantity of carraway powder and ginger. Where there is no fever, the first dose of salts, &c., may be followed up on several successive mornings with four ounces of Epsom salts, two ounces of powdered gentian, and one ounce of powdered ginger, mixed and given as above.

INFLAMMATION OF THE LIVER.

This takes place chiefly in stall-fed cattle.

Symptoms.—Yellowness of the eyes and skin; ordinary symptoms of fever; the animal lies continuously on the right side; there are slight spasms in that side, or wavy motions of the skin over the region of the liver; a general fulness of the belly, chiefly on the right side, and the expression of considerable pain when pressure is made on that side. There is usually some degree of constipation, and the urine is scanty, and of a yellow, or brown, or bloody colour.

Treatment.—Bleeding; physic; blisters on the right side, and restricted diet, from which everything of a stimulating nature is withheld.

JAUNDICE, OR YELLOWS.

Cause.—Obstruction of the passage of the bile from the gall-bladder to the small intestines; also impairment of the functions either of the liver or the bowels.

Symptoms.—Resembles inflammation of the liver (see same), in the yellow colour of the eyes and skin, and the urine; but there is not the affection of the right side peculiar to that disease. In jaundice there is sometimes at first little inconvenience; in other cases there is irritation and fever from the commencement, excessive thirst, suspension of rumination, dulness, loss of appetite, strength, and condition. If the disease is allowed to go on, a scaly eruption appears on the skin, attended with extreme itching, sometimes degenerating into the worst species of mange.

Treatment.—Aperient medicines, such as Epsom salts, 1 lb.; powdered ginger, $\frac{1}{2}$ an ounce; and if the symptoms do not readily give way, a drachm of calomel may also be given in thick gruel. Gamgee recommends that reliance should chiefly be placed on nitre and Epsom salts, in two or four ounce doses daily of either. Turpentine in linseed tea, either alone or combined with aloes, has also been recommended, and clysters or injections have likewise proved of great service.

SPLENIC APOPLEXY.

Causes.—Professor Simonds attributes this disease to unwholesome herbage, bad water, and insufficiency of water, &c.

Symptoms.—The animal affected first refuses to eat; its back becomes arched; it has a difficulty of progression, a staggering gait, a twitching of the muscles. This is followed by paralysis. The animal has a dull, dispirited look; the head hangs down; frothy saliva comes from the mouth; the breathing is laboured and difficult; the pulse is heightened, and becomes tremulous and indistinct ; and colicky pains come on. There is an infusion into the intestinal canal, griping pains, and diarrhœa ; the evacuations and urine become blood-coloured, and the animal falls, and generally dies in convulsions. Some die frantic; others in a state of coma or insensibility. The ordinary duration of the malady is not more than 18 hours, and usually from six to eight hours. If it lasts for 24 hours, Professor Simonds states that a change is wrought in the condition of the blood; and that if the blood thus changed happens to be swallowed by other animals, it produces an effect almost equal to that of prussic acid. He has known it to be thrown into a yard, where it was eaten by pigs, and within a few hours after the pigs died.

Treatment.—Professor Simonds does not put confidence in any particular mode of treatment. All blood affections, to which class splenic apoplexy belongs, are exceedingly fatal, and run their course so quickly that nothing scarcely can be done to arrest them.

DIARRHŒA.

Diarrhœa, or "Scouring," may be simple, or connected with other diseases.

Causes.—Acute diarrhœa may be the result of the abuse of purgatives ; feeding on certain poisonous plants ; sudden change of food, generally from dry to green food ; excess of food; bad water; or some humid and unhealthy state of the atmosphere.

Treatment.—In simple cases a mere change from green to dry food will generally effect a cure. When the disease is more obstinate, first give a mild purgative, such as a pint to a bottle of castor oil, and after the bowels have been cleared out, give the following mixture :—

Prepared chalk ...	...	...	2 ounces.
Powdered catechu	...	...	1 ounce.
Powdered ginger ...	...	...	4 drachms.
Powdered opium ..	...	...	1 drachm.
Peppermint water...	...	...	1 pint.

Mix together, and give in thick gruel. If the disease has assumed a chronic form, give two doses of the oil—one after the other has operated, with ten grains of opinm in each dose, and then give the powder above specified, but with a double quantity of ginger, and half a drachm of powdered gentian in addition. After a time, a drachm of Dover's powder may be given morning and night, and when that ceases to have effect, return to the first powder. Sometimes the liver is affected, and in that case, after a dose of salts or oil has operated, give calomel in combination with opium, half a drachm of each twice a day. Sulphate of copper has been found extremely useful in chronic diarrhœa in cattle, the dose not to exceed two drachms.

Diarrhœa is often very troublesome and fatal when *Calves* are affected by it, particularly those which are hand-fed. The *symptoms* in calves are voracious appetite, langour, abdominal pain, tendency to swelling in the abdomen, and frequent discharge of yellowish white excrement.

Treatment.—Injections; a little chalk or wheaten flour in the milk, and doses of carbonate of magnesia or carbonate of soda, one to two drachms each, according to the age of calf. These last medicines are particularly desirable when there is any tendency to swelling. A tablespoonful common rennet, as used in making cheese, may be given with advantage after the calf has taken a little milk, as the rennet assists the natural action of the stomach, and prevents derangement.

Another useful dose is composed of two drachms of alum, dissolved in a pint of hot milk, to which a drachm of ginger may be added, and a scruple of opium.

DYSENTERY.

This is a severe form of diarrhœa (see "Diarrhœa"); and along with the treatment recommended under that disease, employ injections, and even moderate bleeding. With mild aperients in the first instance, reliance must be placed on the calomel and opium, as described under "Diarrhœa." During recovery give tonics, such as gentian and ginger mixed, one to four drachms of each; and attend to diet, relying on bran and linseed mashes.

RED-WATER.

This disease is locally known as " Moor-ill," " Daru," " Dry Murrain," &c. It affects store as well as dairy stock, and not unfrequently cows after calving.

Causes.—When cattle are seized with it when on pasture, it is generally believed to arise from an abundance of acrid and astringent plants in the pasture. It is peculiar to certain districts, and sometimes even to different fields on the same farm ; cattle grazed on places and fields in the immediate vicinity being perfectly free from any liability to the disease ; while cattle bred in districts subject to it are much less liable to be attacked than cattle brought from a distance. It is an affection of the kidneys, complicated with other derangements of the system.

Symptoms.—A discharge of bloody urine, which becomes black if the disease goes on unchecked ; diarrhœa at first, which suddenly changes into obstinate constipation ; the animal ceases to feed ; the pulse and breathing are quickened, the former being weak, and the extremities become cold. When it attacks milch cows, the yield of milk is materially diminished, and the animal becomes hide-bound, and ceases to ruminate.

Treatment.—The great point in treating red-water is to open the bowels as expeditiously as possible. Give at once a pound of Epsom salts, and same of treacle, and half pound doses of each every six or eight hours afterwards until the bowels are thoroughly acted upon. Aromatics, such as ginger, may be omitted in the first instance ; but should the treatment be protracted, powdered ginger should be administered freely with the after doses. When Epsom salts were not at hand, we have found a bottleful of linseed oil an excellent substitute ; and lacking both, give plenty of common salt, diluted with large

quantities of water. Once the bowels are fairly opened the
danger is past ; but we have sometimes found it exceedingly
difficult to produce purging in cases of red-water. The action of
Epsom salts or oil may be increased by adding from 6 to 10
grains of croton seed to the dose of salts or oil. In order
to produce purging, Mr. Spooner recommends the following
draught :—

Epsom salts	12 ounces.
Sulphur	4 ,,
Carbonate of ammonia	4 drachms.
Powdered ginger	3 ,,
Calomel	1 scruple.

To be made into a draught with warm gruel. One-fourth of the
above may be repeated every six hours, without the calomel, until
the bowels are relaxed ; after which mild stimulants, with diuretics, may be given, such as the following :—

Spirit of nitrous ether	1 ounce.
Sulphate of potash	2 drachms.
Powdered ginger	1 ,,
Gentian	1 ,,

To be given in linseed gruel twice a day. Injections of warm
water are very serviceable before the bowels are relaxed.

INFLAMMATION OF THE KIDNEYS.

Causes.—Cold and wet ; bad food, and blows or strains of the
loins. It is also sometimes brought on by abuse of diuretics, or
by the improper administration of cantharides, &c. It frequently
occurs in the spring season, especially when the animals are
much exposed, and the weather ungenial or stormy.

Symptoms.—Great pain and weakness of the loins ; the urine,
which is highly coloured, is scanty and discharged with difficulty,
and there are general feverish symptoms.

Treatment.—Where there is much fever, bleed in the first instance ; give castor or linseed oil as purgatives ; stimulate the

loins with a mustard poultice, or apply a fresh sheep-skin to the loins, flesh side under; give linseed gruel in large quantities, and administer injections of warm water.

Sometimes blood in a coagulated state, or clotted, is passed with the urine. This is frequently caused by cattle riding on one another, and should be treated as above. Mr. Spooner says the following medicines have often succeeded in such cases, where others have failed:

Oil of turpentine ...	...	...	1 ounce.
Tincture of opium...	...	...	1 „
Oil of juniper ...	...	...	2 drachms.

To be given in linseed gruel. This last named form of disease is called " Hœmaturia."

BLACK-QUARTER.

This disease is also known as " Black-leg," " Quarter-ill," " Blood-striking," &c., and by veterinarians as " Inflammatory Fever," although it lacks certain of the regular train of symptoms which are always present in inflammation.

In black-quarter we find that it is mainly confined to calves between six and eighteen months old; that as a rule it affects the best animals, and those growing fastest; that it generally occurs when they have been changed from a comparatively poor diet to improved food; that ill-growing, lean, unthrifty animals are seldom affected by it. When it occurs in the last named class of animals, the stomach will be found to contain a mass of indigestible matter. Sudden changes of temperature, such as occur when the days are warm and the nights frosty; a cold, rainy night, after a course of mild weather, have also been found to be exciting causes. At the same time, black-quarter also affects animals tied up in the house, and not exposed to any atmospheric influence, in which case the cause is to be looked for amongst those first mentioned.

Black-quarter is a malady of the blood, the chief feature of which is the sudden effusion of that fluid into the tissues, and a tendency to decomposition of the effused fluid. The effusions take place in different organs and localities of different animals: when they occur in the spleen or milt they are commonly called " Splenic Apoplexy" (see same); when the effusion takes place beneath the skin, over the haunch, or one of the limbs, it is designated "black-quarter," &c. The disease also occasionally developes itself in the region of the throat, and is then called "Gorge evil." The rapidity with which it causes death, which it generally does despite all treatment, greatly depends on the organ or place principally attacked. If the effusion is within the skull, death is almost instantaneous, from the pressure of the effused blood on the brain, the ventricles and substance of which are frequently found the seat of blood effusions. Occasionally the blood is found effused in the sac containing the heart. At other times, such as in black-quarter, properly so-called, the principal seat of effusion is beneath the loins or one of the limbs. Although the malady has so many different names, according to the different parts most visibly attacked, and each name is intended to represent a particular disease, different from any other, they all proceed from identically the same diseased state of the system, which Professor Ferguson has designated apoplectic congestion.

The *post-mortem* appearances are worthy of attention in the case of a malady of this nature. There is always excessive congestion and extravasation of blood in loose tissues, such as at the shoulder, on or under the loins, inside and outside of hind leg, &c., and in cases of some duration, where life has been retained for some days, there is sometimes ulceration in the bowels, which, together with the lungs, are one black-jellied mass.

Symptoms.—These are rapid, sometimes astonishingly so, in development. We not unfrequently find a young beast or two dead in the morning, which were apparently quite healthy over night. There is dulness, and generally cessation of rumination and appetite, although in some cases the animals have ruminated to the end. There is a halting or lameness, without any apparent pain, as the animal will sometimes drag the lame limb without appearing to feel it ; at other times it seems as if paralyzed. The part where the disease is seated is commonly marked by more or less swelling, the external character of which is diffused, puffy, and crepitating or crackling, but not invariably so, it being sometimes hard and tense, but seldom circumscribed. Where crepitation is absent the extravasation is more deeply seated than on the surface, and in such cases the appearance of the limb sometimes causes it to be mistaken for the effects of a hurt ; but where there is a severe hurt there is generally increased heat, and other symptoms of fever ; whereas in " Black-quarter" the temperature is below par, especially of the extremities. Where congestion and extravasation take place in the lungs, there is a quickened respiration, moaning, &c., but not much pain. The disease rarely attacks pregnant young animals or aged cows, which is not the case in purely inflammatory disease.

Treatment.—Curative treatment is generally, indeed we may say invariably, useless. We cannot in a moment, nor yet in a sufficiently short time, so change the blood as to cure the disease, and what we see and regard as symptoms are but the precursors of death. Bleeding is of no use, because the vitality of the system is destroyed by extravasation already, and thus bleeding often only hastens death.

Seeing, then, that curative treatment is of little use, we must employ that which is of a preventive nature, and experience has proved that such is within our reach. Among the most effectual is oil-cake. Let calves be accustomed to eat it from an early age, give a little of it to them daily, and gradually increase the quantity as they grow, until by the month of October or November following their daily allowance shall be from 1 lb. to $1\frac{1}{2}$ lb., and by April thereafter let it be increased to 2 lb. Towards the end of autumn put a seton, which has been smeared with turpentine, in the brisket of each calf, and allow it to remain until next spring, giving it an occasional turn, and a fresh smear with turpentine. Previous to or at the time of setoning give each calf six ounces of Epsom salts and one ounce ground ginger. By following this system, accompanied with proper shelter for the animal during the cold months, so few cases will occur that the disease may be considered practically as banished.

It is right to mention that in blood diseases the blood of an animal so affected is capable of producing injurious effects on even the most healthy animals. Many persons have died from the effects produced on their systems by opening, skinning, or dressing cattle affected with the malady.

CHAPTER IV.

PARTURITION, THE UDDER, ETC.

ABORTION, OR SLIPPING CALF.

This usually takes place between the ninth and fifteenth week of pregnancy; and a cow that has slipped or cast calf once will be liable to do so again. When it occurs with one animal in a herd, it usually spreads through the rest of the in-calf cows, from sympathy in some instances, and in others from all the herd being exposed to the same exciting causes.

Causes.—Blows, strains from riding or jumping on other animals, fright, excitement, and from constitutional defects. At same time, there is good reason to believe that it is frequently caused by the growth of ergotised grasses in the pastures. Such are common where the climate is humid, and in rainy summers; and this accounts for the epidemic-like nature of abortion in some years.

Symptoms.—Resemble those of ordinary calving. A regular symptom is the discharge of a red or yellow glairy fluid from the vagina.

Treatment.—The only treatment necessary is to give a dose of salts and ginger (see "Retention of Cleansing"), and afterwards a sedative, such as an ounce of laudanum, and the same quantity of spirit of nitrous ether, mixed. The calf and cleansing should be deeply buried, and the stall disinfected with carbolic acid or chloride of lime. Fatten the cow for the butcher.

INVERSION OF THE WOMB.

The " downfall of the calf-bag," as this is usually called, occurs sometimes in calving, and occasionally even later.

Treatment.—Administer, in the first instance, three ounces of laudanum, partly to quiet the animal, and partly to relax the muscular fibres of the neck of the womb. Put a quantity of litter under the hind feet, so that the hinder parts may stand high. Have the bag well washed with tepid water, to remove dirt, the bag being supported by two persons on a strong, soft cloth. If it has occurred in calving, the cleansing must be gently removed ; then let the cloth be raised until the womb is on a level with the bearing, after which the parts should be returned to their proper place—a process which will require some time and care to effect it. Pass a web collar round the neck of the cow, and a girth of the same material round the body behind the shoulders ; connect this with the collar under the brisket and over the shoulder, and on each side. Another girth is passed a little in front of the udder, and connected with the first in the same way. To this, on each side, and level with the bearing, a piece of stout wrapping cloth or other strong material, 12 or 16 inches wide, is sewed, and brought over the bearing, and attached to the girth on the other side in the same manner. Keep the cow quiet; give her warm mashes and gruel, and if she is restless, small doses of opium.

RETENTION OF THE CLEANSING.

The " after-birth" or "cleansing" *(placenta)*, although usually discharged soon after calving, is sometimes retained, producing irritation and fever. When this is the case, it soon becomes putrid; at the same time, there need not be much fear of the consequence, as it has been retained for a week without injury. It is, however, well to get rid of it, and the best means to use is to give, two or three hours after calving, an aperient drink consisting of a pound of Epsom salts and two drachms of ginger, mixed with a pint of good, warm ale. This will also be useful in other respects. In some tardy cases we have given a quarter of an ounce of ergot of rye in ale, which, acting upon the uterus, caused the speedy discharge of the cleansing, without resorting to mechanical assistance, which should never be employed, if it is possible to avoid it.

INFLAMMATION OF THE UDDER, OR GARGET.

This disease is of frequent occurrence after calving, and when the udder is full of milk.

Causes.—Exposure to cold and wet; leaving the cow for a considerable time without being milked, as we see frequently done when she is brought to market, or exhibited in a show-yard; not milking clean.

Symptoms.—One of the teats or quarters becomes enlarged, hot, and tender; the secretion of milk is interrupted, and then the udder becomes hard and knotty. From the quarter first affected it spreads to others, and there is always a greater or less degree of fever.

Treatment.—If it occurs in the newly calved cow, putting the calf to suck will generally prove the best remedy ; but if this is inadvisable, from other reasons, or cannot be done, apply fomentations of hot water, drying the udder with a linen cloth, after being fomented. Give a full dose of Epsom salts, that is, one lb., mixed with same of treacle, in gruel ; draw the milk gently, but completely, off, at least twice a day ; and if the udder is much swollen, let it be supported by a broad bandage. If the cow shivers, give an ounce of ground ginger, dissolved in warm gruel or ale, with two ounces of spirits of nitrous ether. After fomentation, rub the following ointment into the part :—

Camphor, powdered	...	...	1 ounce.
Mercurial ointment	...	...	2 drachms.
Lard ...	...	...	8 ounces.

To be well incorporated. Wash this off with hot water before every milking, drying well after washing, and after the milking rub on some more of the ointment. Should this treatment not succeed—which it will do if the disease is not allowed to proceed too far—then the following ointment must be substituted for the above, and well rubbed on the part :—

Iodide of potassium	...	...	1 part.
Lard ...	...	...	7 parts.

To be well incorporated. At the same time, the iodide of potassium may be given internally in doses, beginning with six and gradually increased to twelve grains daily.

MILK FEVER.

This is sometimes designated " Puerperal Fever," and sometimes "Dropping after Calving." It attacks cows within the first ten days after calving, and the mortality from it is greatest amongst cows that are highly kept. It is changeable in its character. By some it is considered an affection of the nervous system of the brain and spinal marrow, but principally of the latter, at the region of the loins, and that it is the effect of the great exhaustion

H

of the nervous system produced by parturition and its consequences.   In some cases it appears to be a blood disease, assuming the character of typhus.

Symptoms.—Wild look, staggering gait, the animal lies stretched
out on her side in a state of torpor, and only showing signs of
distress by dismal moanings.   If the head is turned round to the
side, there it remains, and when gruel is poured into her it runs
out of her mouth without any attempt to swallow it.   There is no
appetite, and no discharge of dung or urine.   The pulse is weak,
quick, and often imperceptible.   The disease sometimes makes
rapid progress, at other times the animal may linger one or two
days.   There is a form of the disease in which the symptoms are
similar, but of a milder character.

    Treatment.—Give the following dose of physic :—

Epsom salts ...	...	... 1 lb.
Flowers of sulphur	...	... 4 ozs.
Croton oil ...	...	... 10 drops.
Carbonate of ammonia	...	... 4 drachms.
Powdered ginger	...	... 4 drachms.
Spirit of nitrous ether	...	... 1 oz.

This should be carefully mixed and dissolved in warm gruel, and
given to the cow slowly and carefully.   In particularly severe
cases, and where there is obstinate constipation, the croton oil
may be increased, and from four to eight grains of powdered cantharides may also be added.   A strong stimulating or blistering
liniment should be rubbed on the loins, and a fresh sheepskin
afterwards placed on the loins, the woolly side outwards.   Every
six hours one fourth of the above medicine, with the exception of
the croton oil, should be given, until purging is produced, which
may be facilitated by injections.   If the cow cannot or does not
pass her urine, it should be removed from her by means of the
catheter.   Plenty of nourishing gruel should be given, and bran
mashes with some malt mixed through them.

    We are particularly desirous to direct attention to the following description of a mode of treating milk fever in cows, which was adopted by Mr. Henry Woods, Merton, Thetford, on Lord Walsingham's Home Farm. Mr. Wood's account of the case first appeared in *Bell's Messenger* (1869), and subsequently other correspondents of that journal stated that they, too, had found it efficacious. Mr. Wood says—

"On Saturday morning, the 17th April last, at six o'clock, an Alderney cow, in fair condition, calved her sixth calf, and appeared to go on very well until about six o'clock at night, when the herdsman thought she appeared rather dull. On Sunday morning, the 18th, she was evidently very unwell, was extremely restless, breathed quickly, and would not eat. An opening drink was given, but she still got worse, and on Monday morning, the 19th, at five o'clock, she had lost all her milk, of which she had an abundance the previous night, when she was well and cleanly milked.

" She trembled a good deal and began to stagger, and threw her head about from side to side. About half-past six she fell, and threw her head on to her left side and appeared unconscious. Her bowels were much constipated. At eight o'clock she began to pass off a white, frothy discharge from the nose and mouth. This discharge in previous cases of milk fever that have occurred here generally came on a few hours before death, and I have never before seen a cow recover after such discharge. The herdsman and others who saw the cow were sure she would die in a few hours; but seeing she was perfectly insensible to pain (for when the finger was pressed on the ball of the eye she showed no appearance of feeling it), I thought I might try the effect of powerful stimulants without cruelty to the animal, and determined to do so. With little or no hope of saving her life, I carried out the following treatment :—A drink was given with much difficulty at half-past eight o'clock, composed of one ounce of aromatic spirits of ammonia, three ounces of spirits of turpentine, half a pint of good brandy, and two pints of strong ale. At ten some good oatmeal gruel, with two pints of ale, were given; and as the bowels were very much constipated, a pint of raw linseed oil was administered at half-past ten. At twelve the animal was no worse, the discharge from the nose and mouth had begun to subside, and a second drink, made exactly like the first, was given. At half-past one she had a second pint of linseed oil, and at half-past two more gruel and another quart of ale. At half-past four the cow was found with her head up, and clearly sensible; for she knew the herdsman when he opened the door of her house. There was then no discharge from her nose and mouth. A third drink, prepared as before, was given: and at half-past six she had a third pint of oil, and also some gruel.

"I am quite prepared for many of your practical readers shaking their heads and being incredulous, when I tell them that at half-past eight the herdsman *found the cow on her legs.* She had then a fourth drink, which was repeated every four hours during the night, with gruel and beer given between the doses.

"On Tuesday morning, the 20th, the cow (to the surprise of all who saw her on the previous day) was very much better, and began to feed a little. Her bowels were now nicely acted upon, and the milk began to return. She had a drink at six on Tuesday morning; and, when I found her so much better, I was afraid to go on with such stimulating drinks, and ordered them to be discontinued during the day. She had a drink, however, again in the evening, when she appeared to be going on favourably, and was then left for the night.

"On Wednesday morning, the 21st, I had to leave home early for a few days, and the herdsman came to report that the cow was not so well, and that she would not eat, and moreover she began to tremble again, and had less milk than on the previous day. I then ordered that the drinks, in the same proportions as given at first, should be again given every four hours until Thursday morning, the 22nd.

"It was reported to me that after the first dose on the

Wednesday morning the trembling passed off, and the cow appeared better, and continued to improve. On Thursday, the 22nd, the medicine and gruel were discontinued, and the cow fed nicely, care being taken to give only small quantities of food at a time.

"On Friday, the 23rd, she was turned into a sheltered paddock and ate some grass, and she was out the greater part of the following day, and also on the Sunday.

"On Monday, the 26th, all who saw the cow declared her perfectly well, and so she remained ever since, and is now to be seen a living example of the good effect of stimulating treatment in the case of milk fever.

"I have given the whole case as clearly as I can; and if any person who has a case of milk fever tries the treatment which did so much good here, I shall be very glad if he will tell me the result either privately or through your paper.

"I have seen many cases of milk fever; and so fatal is the disease that I have known but few recover, and when the disease was so far advanced as in the case reported above, I have never before seen a recovery.

"I believe the practice of depletion to be a mistake, as it is now acknowledged to be a mistake in cases of fever in human beings. Two or three years ago a case of milk fever occurred in an Alderney cow on the Merton farm, though in that case the animal was not so far gone as to lose consciousness. She was bled, and died in less than half an hour afterwards. On the other hand, in the case I have described it seems to be proved that the free use of stimulants saved the animal's life; and when they were only given twice on the Tuesday and were omitted during the following night there was a relapse; but when they were again given regularly, night and day, the animal improved in a very marked manner."

Preventive Treatment.—Let the cow have quiet exercise daily before calving, unless the weather is unfavourable. If the cow is full of flesh, give her a half dose of physic some time before she is expected to calve, such as one-half of the purgative mixture recommended above, omitting the croton oil, with bran mashes, and after calving has taken place, the same half dose may be repeated with bran mashes as before, and then care, shelter, and more liberal feeding may follow.

Youatt recommended bleeding in treating this disease, but later veterinarians are against it.

SORE TEATS.

This, like "Garget" (which see), occurs after calving.

Symptmos.—Cracks and sores on the teats, causing much uneasiness when the cow is being milked.

Treatment.—Foment the teats with warm water, taking care to dry them well afterwards, and after milking, dress the teats with the following ointment:—Take an ounce of yellow wax and three ounces of lard ; melt them together, and when they begin to get cool, well rub in a quarter of an ounce of sugar of lead, and a drachm of alum, finely powdered. If there is any obstruction in the teat, insert the small silver tube which is made for the purpose. It is left in the teat, and is so constructed that it can be retained in its place by means of a bit of tape. We have repeatedly found great advantage from the use of it.

COW-POX.

This disease is not of frequent occurrence.

Symptmos.—Numerous pustules on the udder and teats, the contents of which are infectious, and may be propagated by the hands of the milker of one cow to another.

Treatment.—Give a dose of Epsom salts, and apply to the sores on the teats a little powdered chalk, with one-fourth part of powdered alum. Cow-pox is a mild disease, and runs its course without any serious consequences. It is very likely that the attendants will catch the infection.

NAVEL-ILL.

If the navel string of calves continues to bleed, pass a tight string round it, not quite close to the belly, and put a pledget of tow dipped in friar's balsam over the sore spot; confine this with a bandage, and change it every morning and night.

If in spite of this treatment inflammation occurs, foment the part with warm water, and if the tumour points, or comes to a head, open it with a lancet, and give two or three two-ounce doses of castor oil beat up with an egg. Navel-ill is frequently fatal.

BLOODY MILK.

It sometimes happens that more or less blood mingles with the milk, although the cow is in apparently good health.

Treatment.—Bleed moderately, and give a dose of castor oil or linseed oil, say a pint, mixed up with the yolk of an egg; and rub the udder night and morning with some elder ointment. This soothing ointment is made by boiling the leaves of the elder bush in lard.

DRYING A COW'S MILK.

Although not a disease, yet as the means to be adopted for drying a cow of her milk form a frequent subject of inquiry, we may say that the plan to follow is, first to milk her out to the last drop, then bleed her pretty freely, and give a drench composed of 4 ounces of roche alum and 4 ounces of common alum, powdered, in a quart of cold skim milk. Do not milk her again, but keep her on low, dry keep for some days. It is, however, doing violence to nature.

If the cow is again milked, the drench must be repeated. In case of any swelling of the udder or teats, see " Inflammation of the Udder," page 96, and " Sore Teats," page 102.

CHAPTER V.

THE SKIN.

ANGLEBERRIES OR WARTS.

These excrescences vary in size, and are often exceedingly troublesome.

Treatment.—If the warts are small, rub them well with camphorated oil, or they may be removed by touching them daily with nitrate of silver (lunar caustic), and if numerous, by washing them with a strong solution of it, but the knife and hot iron are the most effectual remedy. When they are attached to the skin by a neck they may be removed by tying a waxed silk thread tightly on the neck of the angleberry.

LICE.

Cause.—Poverty of living, and want of attention to cleanliness.

Treatment.—Train oil or linseed oil well and frequently rubbed into the skin is an effectual cure. The ointment recommended for "mange" is also useful. Cleanliness. An infusion of white hellebore used as a wash over the body has been recommended.

MANGE.

This is a very contagious malady, which chiefly affects the back and neck, and about the tail.

Causes.—Parasites on the skin; contact with affected animals; want of attention to cleanliness; poverty.

Symptoms.—An intolerable itching; the affected part becomes sore, the hair falls off, and the skin becomes thickened, and drawn up in folds or wrinkles.

Treatment.—Wash well with soap and water, using much friction in drying the washed parts. Then rub well in the following ointment :—

Flowers of sulphur ...	...	4 ounces.
Linseed oil or train oil ...	...	8 ,,
Oil of tar or oil of turpentine	...	2 ,,

To be thoroughly mixed before using.

RINGWORM.

This disease prevails chiefly among calves and young cattle.

Causes.—Confinement in close, ill-ventilated premises; indifferent feeding ; contact with affected animals.

Symptoms.—A scurfy eruption, accompanied by redness and itching, which is speedily followed by shedding of the hair. It usually affects the eyelids, face, cheeks, and shoulders.

Treatment.—The ringworm spots should be rubbed daily with a mixture of equal parts of iodine ointment and sulphur ointment, or use the ointment of the iodide of sulphur, which should be made up only as required, and used immediately afterwards. The spots may also be lightly touched with a stick of lunar caustic. The oxide of zinc ointment is also an excellent remedy when rubbed in daily on the parts affected.

SPRING ERUPTION IN THE SKIN.

Causes.—Change of keep, especially when cows have been poorly kept during the winter. Change of coat.

Symptoms.—Much irritation of the skin ; the surface of the body, and even the limbs, become covered with inflammatory spots ; hard pimples form, which are broken by the rubbing, and a scab or crust forms. In a few days the scabs drop off, and the skin remains without hair.

Treatment.—A dose of Epsom salts with some powdered ginger, given in gruel. Wash the skin thoroughly with soft soap and

water.   Attend to cleanliness, and give nourishing diet of a cooling nature.

WARBLES.

Causes.—Eggs deposited in the skin by the cattle fly in summer.

Symptoms.—Small tumours in the back, &c., varying in size
from that of a hazel nut to a walnut.

Treatment.—Forcibly squeeze each tumour, and the maggot
which it contains will be ejected. No further treatment necessary.

HIDEBOUND.

This is usually a symptom of fever or some other disease ; and
coarse-bred cattle are, for the most part, hidebound when in a
lean state.

Treatment.—If unconnected with other disease, it will yield to
the action of purgative medicine and nourishing food.

CHAPTER VI.

MUSCULAR SYSTEM AND EXTREMITIES.

RHEUMATISM.

This disease is known in some districts under the names of
" Chine Fellon" and " Joint Fellon."

Causes.—Exposure to cold and wet, particularly after calving,

or too soon after recovery from serious illness. It is chiefly prevalent in a cold, marshy country; in places exposed to the coldest
winds; in spring and autumn, when there are the greatest vicissitudes
of heat and cold; and in animals that have been debilitated by
insufficient diet, and that cannot, for that reason, withstand the
influence of sudden changes of temperature.

Symptoms.—Great stiffness and pain in moving, attended with
considerable fever. When the animal is urged to move, there is
a marked stiffness in the action, and at first the disease appears
confined to the back and loins, as the animal shrinks when pressed
on the loins; but the stiffness gradually extends to the fore or
hind limbs; some of the joints swell, become hot and tender, and
the animal can scarcely bend them; every movement being
accomplished with difficulty and evident pain.

Treatment.—The treatment must chiefly consist in making the
animal comfortable, and in sheltering from the cause of complaint.
Give purgative medicine—1 lb. of Epsom salts and 2 oz. of ground
ginger, in a quart of warm ale or gruel; and after the bowels have
been well opened, give a dose or two of sulphur, the dose of
which varies from 6 to 12 ounces, according to the size of the
animal. The parts affected should be fomented with hot water,
and then well rubbed with a stimulating liniment, such as camphorated oil, or spirit of turpentine and laudanum. The following is also a useful embrocation, to be applied with much friction:
—Take oil of turpentine, tincture of opium, soap liniment, of
each one ounce; tincture of capsicum, one drachm; mix. A bag
of bran soaked in hot water and covered with a rug may also be
applied to the loins, but it must not be left till it gets cold.

PARALYSIS, OR PALSY.

Rheumatism (see "Rheumatism") is closely akin to paralysis,
and frequently ends in it.

Causes.—Exposure and poor keep.

Symptoms.—Inability to rise; tenderness upon the loins and about the rump, and tightness of skin covering these parts. The animal is also unable to lift its tail in dunging or making urine, hence the idea of "Tail-ill," "tail-slip," or "tail-rot," as a distinct disease, supposed to arise from the existence of "a worm" in the tail. Paralysis chiefly occurs in milch cows and young beasts.

Treatment.—The first thing to be attended to is comfort. Put the affected animal in a warm, but not close, cow-house; give plenty of litter, and put a bag of hot bran, as above, and a warm rug over the loins and back. Keep the bowels open by means of a dose of Epsom salts—$\frac{1}{2}$ lb. to 1 lb., according to the size of the animal—with half an ounce to an ounce of powdered ginger. Give in a pint of warm ale. Hand-rub the loins well, two or three times every day, applying at same time a stimulating lotion composed of equal parts of spirit of turpentine, camphorated spirit, and hartshorn.

JOINT-EVIL.

This disease occurs in the bones at the joints, and is accompanied by pain in the part and fever; there is inflammation and effusion in most cases; a swelling of the bone, which becomes porous, and the part or joint is apt to become bent. It sometimes follows chronic and repeated rheumatism, and it is occasionally the result of injuries. It is frequently seen in very highly-bred and pampered calves; and it is also seen where calves are kept in damp, dark places, and not allowed sufficient exercise and nourishing food.

Treatment.—Give plenty of good diet, especially milk, if the calf is very young. Give also plenty of chalk to lick, and soft, dry bedding.

The following liniment may be applied to the joints and well rubbed in:—Iodine, $\frac{1}{2}$ ounce; glycerine, 2 oz.; mercurial ointment, 2 oz.; olive oil, 6 oz.; mix.

FOUL IN THE FOOT.

This disease is of the nature of foot-rot in sheep.

Cause.—Pasturing on marshy land.

Symptoms.—Inflammation, lameness, and soreness between the claws, and the discharge of offensive matter from the foot.   Abscesses sometimes form again and again, and prove extremely troublesome.

Treatment.—Remove the animal to a place where the feet can be kept dry. A large pledget of tow, covered with tar, on which sulphate of copper may be spread, should be applied between the claws, and renewed every second day.   Applying at first lunar caustic, and following up with tincture of myrrh, treatment which has been most successful in foot-rot, will also likely be useful in this disease ; also the application with a feather between the claws of a mixture of one part of carbolic acid and four parts of glycerine ; or the ointment of carbolic acid on a pledget of tow.   The application of a linseed poultice to the foot, in the first instance, is often advisable.   The poultice may be kept in its place by a stout linen cloth, having two holes in it to allow the hoofs to pass through.

CHAPTER VII.

SPECIAL DISEASES.

~~~~~~~~~~~~~~~~~~~~~~~~~~~

FOOT-AND-MOUTH DISEASE.

This malady may be termed the most contagious, and yet the mildest, of the epizootic diseases which affect the ox.

Symptoms.—A person who is practically familiar with the diseases of stock will at once recognise this affection by the gait and general aspect of the animal: lameness, discharge of saliva, and a peculiar smacking of the lips are indications which are quite unmistakable; but the observer will also distinguish other and not less characteristic phenomena. At the commencement of the affection the animal is dull and inclined to stand still, with the head somewhat depressed and the back arched. In some cases the hind feet are snatched up suddenly, and shaken as though to get rid of something which annoys. The appetite is impaired; there is a discharge of saliva, and also an increased secretion of tears, which trickle down the sides of the face; the milk is diminished, but only to a slight extent at first. The internal temperature at this early period varies considerably, and the test of the thermometer is only valuable as an indication of the amount of fever, and the severity, therefore, of the attack. A comparison of many observations has led to the inference that the average temperature is 104 degs.; but the range is from 102 degs. to 107 degs.

In a few hours after the manifestation of symptoms of illness, vesicles filled with a limpid fluid appear on the inside of the top lip, on the upper part of the tongue, on the palate, on the teats, between the claws, and often on the outside of the coronet just above the hoof, and sometimes on the soft parts of the heels. Those in the mouth are generally few in number, and of large size; their situation is on the top of the tongue and on the palate, as far back in many instances as the soft palate. An examination of many heads of diseased cattle which were slaughtered has quite established the fact of the frequent existence of abrasions on the palate, bearing a general resemblance to those found in cattle plague.

When mouth-and-foot complaint prevails in a mild form, the morbid action subsides in a few days; the vesicles burst and discharge their contents; the exposed surface becomes covered with yellow exudation, and the abrasion is quickly healed; the lameness ceases; the milk, which had been lessened to the extent of, perhaps, one-third less than the ordinary quantity, returns; the animal regains its appetite, and soon recovers the lost condition.

Between this mild form of the disease and the most virulent, there are an infinite number of grades, which depend upon constitutional peculiarities and variation of surrounding circumstances. Sometimes the bursting of the vesicles is followed by ulceration and extensive loss of structure. The hoofs slough off, and even the ligaments of joints in the vicinity of the foot become disconnected from the bones, causing open joints. Deep abscesses form in the mammary gland, the secretion of milk is almost or even entirely suppressed, and the acute disease degenerates into a low fever, which continues for a long time, inducing debility and extreme emaciation, and in some cases ending in death.

Complication of the disease with other maladies is not uncommon, and the fatality which is ascribed in certain instances to mouth-and-foot complaint is often in reality due to pleuro-pneumonia, congestion of lungs, or disease of the digestive organs, any of which may exist in connection with it.

In ordinary instances the duration of the attack is not more than a week; but after the disease has begun to decline there is a period of convalescence of some length, during which the animal suffers from debility. When the appetite is quite restored, the improvement is very rapid, and, as in the case of febrile diseases generally, the recovered animals thrive more than the healthy beasts which are placed under similar circumstances of feeding and management. A second attack of the disease is not an uncommon occurrence.

Treatment.—At first all medicine should be withheld; the animal should be supplied with soft food, succulent grass, pulped roots, mashes of bran and linseed, or thick gruel. No attempt, unless under competent direction, should be made to open the mouth, or to horn down any fluids, alimentary or medicinal. If the fever is very severe, an ounce of nitrate or bicarbonate of potass may be dissolved in the drink water, and a solution of the first-named agent, 1 part to 40 parts of water, may be syringed over the feet frequently, for the purpose of keeping them clean and cool. Bleeding and purgatives are absolutely deadly; even mild laxatives, which do no harm, are of questionable utility; and the use of powerful astringent, styptic, and caustic lotions to the sore mouth is to be carefully avoided.

As an antiseptic remedy, hyposulphite of soda has been much advocated in this and other infectious diseases, and has the positive merit of being innocuous; the administration of 4 to 8 ounces daily of this neutral salt in the drink water may advantageously take the place of more active measures. We cannot, however, speak in its favour as a preventive of infection.

Under the system of treatment suggested, and which is of the least obtrusive character, the sick animals, in the majority of instances, will progress favourably towards recovery. Under unfavourable circumstances of locality, atmospheric conditions, or state of constitution, the malady may assume a virulent type, and in that event the aid of the scientific veterinarian is required. Sloughing of tissue will be met by the use of antiseptic lotions, among which solutions of chloride of zinc, carbolic acid, and tannic acid are most effective. Carbolic acid or chloride of zinc is best adapted for application to the feet, and tannic acid for the mouth. A solution of one ounce of tannin to a gallon of water will be suffi- ciently strong ; and a little of this fluid poured gently into the mouth, and allowed to run to the back of the tongue, will allay irritation and protect the excoriated surface by coagulating the albumen of the secretions.

Tonics and stimulants are necessary when the disease assumes a low form, and they are best administered mixed with the food. Salts of iron, stout, and brandy may be employed with advantage in these cases ; but under good management at the commencement, it will rarely be necessary to give any other tonics than properly prepared and nutritious food.

RINDERPEST, OR CATTLE PLAGUE.

This is what may be called an Act of Parliament disease, being, like pleuro-pneumonia, mouth-and-foot disease, &c., so far as the British islands are concerned, the fruits of unrestricted trade in live stock from the continent, especially from Dutch and Belgian ports.

Symptoms.—Mistakes have sometimes been made between "Rinderpest" and other forms of disease, such as pleuro-pneu- monia and mouth-and-foot disease. The early symptoms of the plague are usually a remarkably dull and dispirited condition of the animal, which will stand with his head hanging down, ears drawn back, and coat staring, and refusing all food, and occasion- ally shivering. The eyes have an unusual expression of anxiety, and a mucous discharge flows from them, and also from the nos- trils. The skin is hot, but sometimes chilly ; the temperature varying from time to time. The extremities are cold ; the breath- ing short and thick, and frequently accompanied with moanings indicative of pain. The inner part of the upper lip and roof of the mouth is reddened, and often covered with raw-looking

I

spots. The bowels are occasionally constipated, but in most instances diarrhœa soon sets in, the evacuations being slimy and very frequently of a dirty yellow colour. The vagina is often intensely reddened. The prostration of strength is great, the animal staggering when made to move. In milch cows the secretion of milk is rapidly diminished, and soon ceases altogether.

Treatment.—Immediate slaughter ; bury the carcass without skinning, several feet deep ; cover it with hot lime ; bury also all dung, litter, &c., and use freely on woodwork, and in stalls, &c., carbolic acid, one gallon mixed with ten gallons of water, as a disinfectant.

Preventive Treatment.—The exclusion of all animals from continental countries—except Spain and Portugal, where the disease has never appeared—except such as are fit and intended for immediate slaughter, and the slaughter of such at the port of debarkation.

Aromatic Tonic for Cattle.—Equal parts of coriander and carraway seeds, finely bruised ; one ounce of the mixture to be given with a tablespoonful of salt, in a wet mash.

Bitter Tonic.—Half an ounce of gentian and half an ounce of ginger, both powdered, to be given in warm beer.

PART III.

THE DISEASES OF SHEEP.

CHAPTER I.

THE HEAD AND NERVOUS SYSTEM.

APOPLEXY.

Cause.—A sudden determination of blood to the head, from over luxuriant pasture, &c.

Symptoms.—Dulness, unconsciousness, redness of the membranes, and partial or even total blindness.

Treatment.—Bleed freely from the neck vein, and give two onnces Epsom salts. The dose for lambs must be less in proportion.

INFLAMMATION OF THE BRAIN.

Causes.—Sudden change from poor to excess of rich food.

Symptoms.—Extreme violence, which in lambs assumes such a form, Mr. Spooner says, as to induce ignorant people to believe they are bewitched.

Treatment.—Same as for apoplexy.

BLINDNESS.

Blindness in sheep usually accompanies or precedes other maladies, or it may be of the nature of ophthalmia, caused by cold, exposure, &c.

Treatment.—When the eyelids adhere, they should be gently separated by means of a bit of well-polished hard wood, like a small paper knife; then bathe the eye with lukewarm water into which some landauum has been dropped, and rub the lids with a little oil, to prevent renewed adhesion. It is very common in cases of partial blindness to blow some powdered glass, alum, or some similar gritty substance into the eye; but this should never be done : it ruins the eye. The following are some simple washes, which may be used in affections of the eye :—

Sugar of lead	...	...	$\frac{1}{2}$ drachm.
Opium wine	...	...	$\frac{1}{2}$,,
Water	...	...	1 pound.

Or,

Goulard's extract	...	$\frac{1}{2}$ ounce
Extract of belladonna	...	1 scruple.
Distilled water ...	...	1 pint.

Or,

Tincture of opium	...	$\frac{1}{2}$ ounce.
Distilled water ...	...	8 ounces.

Or,

Sulphate of zinc...	...	...	1 drachm.
Opium wine	...	...	$\frac{1}{2}$,,
Water	...	...	1 pound.

Or,

Tincture of opium	...	...	$\frac{1}{2}$ drachm.
,, of myrrh	...	...	$\frac{1}{2}$,,
,, of saffron	...	...	$\frac{1}{2}$ ounce.
Rectified spirits of wine	...	...	$3\frac{1}{2}$ ounces.

To be used where there is much inflammation.

BLACK MUZZLE.

This is an eruption which sometimes occurs on the face and nose, and is supposed to be caused by the acrid nature of certain plants.

Treatment.—Apply the following ointment :—

Sulphate of zinc, powdered	..	4 drachms.
Alum, powdered	...	4 ,,
Lard	...	1 lb.

To be well rubbed down together.

The following paragraph is taken from the report, for 1869, of the Governors of the Royal Veterinary College :—

" Some interesting cases of disease of the skin of lambs were brought to the notice of the students. The disease possessed all the characteristics of the affection known as *Crusta lactea* in the human infant. The parts principally affected by the morbid action were the sides of the neck ; and even the shoulders were ultimately attacked. Thick crusts of a dark colour covered the skin, which was also much inflamed and cracked. The young animals suffered much from local irritation and symptomatic fever, under which some sank. The cases, however, were not numerous in the several flocks in which the disease appeared—not more than six or eight among 150 to 200 animals. Application of the oxide of zinc ointment proved beneficial ; but careful nursing and protection both from hot and wet weather were needed as adjuncts to the treatment."

EPILEPSY.

Mr. Spooner says that this disease, which is an irregular action of the nervous system, occurs mostly with young sheep in good condition, when they are turned into a pasture early in the morning, whilst the hoar frost is on the ground. The paunch becomes chilled, and blood is determined to other parts.

Symptoms.—Convulsive fits, during which the sheep will stagger, reel, and fall.

Treatment.—Avoid the cause, and give a little hay or cake, or equal parts of cake and crushed grain in the morning.

GID, OR STURDY.

Causes.—This disease is caused by the existence of hydatids on the brain. These hydatids are produced by means of eggs taken with the food, and deposited in the brain by the circulation of the blood. They appear like watery bladders, varying in size from a small nut to that of a hen's egg. Exposure to long continued wet, combined with poor feeding, aggravates the disease. It is chiefly confined to hoggs, and it is not an infectious disease.

Symptoms.—The sheep appears dull and moping, keeps separate from the rest of the flock, keeps its head high and a little to one side, is unsteady in its movements, and after a time moves in a circular manner; there is a partial or total blindness, and the eyes are blue and wandering.

Treatment.—On examining the skull a part softer than the rest will be found; cut an incision round the soft part, leaving a small attachment as a hinge, lift up the part so cut, and carefully remove with a shoemaker's awl the bladder, without breaking it; then replace the cut portion of skull, cover it with a plaster of tar on a piece of linen, and put a cap over all. The animal may recover after this treatment, but in most cases it is better to slaughter the animal at once when it is observed to be affected, as the flesh may be used as food.

LOCKED JAW.

Causes.—Exposure to cold and wet, and also from castration performed in a rough, bungling manner.

Symptoms.—A peculiar motion of the head; fixed jaw; rigid neck; grinding of the teeth; rigidity of the muscles.

Treatment.—This disease rapidly runs its course. Give external warmth; also aperient medicine—2 oz. Epsom salts and 2 drachms ginger—followed by 2 drachms tincture of opium, with the same of ginger, given in gruel twice a day ; proportionate quantity for a young lamb.

LOUPING-ILL, OR TREMBLING,

Attacks sheep and lambs in spring, especially during the prevalence of dry, frosty, and easterly winds.

Symptoms.—Dulness, deadness of coat; loss of power of one or both sides ; tremblings ; contractions of the gullet ; convulsive fits ; disturbed breathing ; gnashing of teeth ; and foaming of the mouth ; and a sidelong motion of the body.

Treatment.—Copious bleeding from the neck, followed by purgative medicines, such as Epsom salts, 1 to 2 ounces ; camphor, $\frac{1}{4}$ to $\frac{1}{2}$ drachm. Give shelter.

PALSY.

Cause.—Cold and moisture. It more frequently affects lambs than older sheep.

Symptoms.—The powers of the nervous system are suspended, either wholly or partially, and the loins are more frequently affected than other parts.

Treatment.—The object of treatment is to promote warmth externally and internally, and with that view give the following :—

Powdered gentian	1 drachm.
„ ginger	1 „
Spirit of nitrous ether ...	2 „

to be given in gruel twice a day to a sheep, and one-fourth or one-half of the dose to a lamb.

RABIES.

Cause.—Bite of a mad dog.

Symptoms.—These show from two weeks to six weeks after the injury, and are as follow :—Propensity to mischief, furious butting, diminished appetite, disposition to ride other sheep, nervous irritability, twitching of the muscles, quickened respiration, occasional drowsiness, thirst, but with a difficulty in swallowing, and a flow of saliva from the mouth, which is dangerous to persons handling the sheep.

Treatment.—Hopeless; slaughter the animal; bury the carcass deep in the ground without skinning it, and cover it with hot lime.

CHAPTER II.

DISEASES OF THE RESPIRATORY ORGANS.

BRONCHITIS.

Cause.—Worms in the windpipe.

Symptoms.—Similar to the disease in cattle.

Treatment.—Half a pint of lime water for a sheep, and a quarter of a pint for a lamb, should be given morning and evening, or two tablespoonfuls of salt dissolved in water, continuing the treatment for a week.

CATARRH.

Causes.—Unduly wet seasons, combined with exposure and bad general treatment.

Treatment.—Good nursing and shelter will frequently remove catarrh, without any other treatment. If the cough is severe, give half an ounce of Epsom salts, and a drachm each of nitre and of ginger, and half a drachm of tartarized antimony, dissolved in gruel. In very severe cases bleed from the neck.

PNEUMONIA, OR INFLAMMATION OF THE LUNGS.

This disease is allied to pleurisy, which is frequently combined with it.

Causes.—Anything which chills the system ; consequently, not uncommon after washing.

Treatment.—Copious bleeding from the neck; purgative medicine, such as Epsom salts, 2 ounces ; after which give the following mixture, as a daily dose for a sheep:—

Nitrate of potash	1 drachm.
Tartarized antimony	10 grains.
Ipecacuanha	5 ,,

Mix. Insert setons in the brisket. Similar treatment to be followed in Pleurisy.

CHAPTER III.

DISEASES OF THE INTERNAL ORGANS.

SPLENIC APOPLEXY.

See " 'Splenic Apoplexy" in cattle.

Treatment.—When the malady has declared itself it is certainly fatal, and medicines are of no avail; but the rest of the flock should be turned at once on scanty pasture, allowed the free use of rock salt, and each should get a dose of two ounces of Epsom salts and 2 drachms powdered ginger in some warm ale or water.

INFLAMMATION OF THE BLADDER

Occurs chiefly in rams fed liberally on cake and beans, but is not a common disease.

Symptoms.—Disposition to stale frequently, with inability to pass the urine.

Treatment.—Bleed freely from the neck; give 2 ounces Epsom salts and 3 grains opium in warm gruel; give warm water or linseed tea both by the mouth and as injections, and with it a drachm of laudanum two or three times a day.

URINARY DISORDERS AMONGST FATTENING SHEEP.

The following remarks on this subject are from the pen of the veterinary editor of the *North British Agriculturist :—*

The liberal dietary and artificial forcing of sheep now so commonly pursued on all well-cultivated farms yield to the farm a fair return in meat, wool, and manure ; but such management unfortunately entails losses from disease, from which unimproved and less carefully-tended flocks were a few years ago exempt.   Complaints of the prevalence of tapeworm and other entozoa are greatly on the increase wherever sheep have been largely multiplied. Fatal congestion of the various internal organs cuts down scores of well-bred, good-thriving lambs. In some of the best cultivated counties of England, flockmasters during the spring months have been much troubled by their sheep dying from bladder and urinary disorders. In Oxfordshire and some of the adjoining counties of England several breeders have this spring [1870] experienced considerable losses from these urinary complaints. Ewes and ewe hoggs, with their direct and straight canal from the bladder to the external parts, are not subject to this urinary complaint. Rams and wethers of all ages suffer, owing to the peculiar formation of the urethral canal, which not only curves round the pubis immediately after leaving the bladder, but as it approaches the sheath is reflected upon itself, constituting what is called the worm or vermiform appendage. At this point of its tortuous course the tube is also narrower than elsewhere, and hence the extreme liability of its being here blocked up by any sedimentary matters which happen to be contained in the acid-concentrated urine. When ram or wether sheep are freely fed on rich nitrogenous diet, have little water and little exercise, the urinary bladder and the duct which empties it—technically called the urethra—become loaded with slimy mucus and gritty sedimentary matter, which thus gradually block up the passages. The urine, although regularly and abundantly secreted, fails to escape; its retention causes the animal much annoyance, and in the large proportion of cases destroys life. Three years ago this troublesome disorder was sensibly and lucidly described by Mr. Litt, the well-known veterinarian of Shrewsbury, in a prize report published in Vol. III. (second series) of the "Journal of the Royal Agricultural Society of England." The better kept the sheep, the greater the liability to this urinary complaint. A liberal supply of mangel appears to aggravate the evil, probably on account of the saccharine and saline matters so abundantly contained in the mangel inducing congestion of the kidneys, and thus after a time rendering the urine more turbid and apt to deposit in the passages its solid consti- tuents.

The earlier symptoms of acid, high coloured urine dribbling away at unduly long intervals, usually escape notice. The affected sheep is first observed to be careless about his food; he does not come up to the trough with his fellows; he stretches his hind legs as if desirous to stale, but without effecting his purpose; he be- comes dull; by-and-bye he gets restless, and strains frequently; he pants and refuses his food. If turned up, the point of his sheath, instead of being perfectly dry, as in health, is found to be moist, and occasionally the wool along the belly is wet from the urine trickling away whilst the animal is lying down. From the arch of the pubis to the point of the sheath, the urethra may generally be felt choked up and distended with thick, yellowish- white sediment, which on analysis is found to consist mainly of ammonio-magnesian phosphate. The symptoms often continue for two or three days, when the animal sinks exhausted and poi- soned by the urinous materials which he is unable to get rid of. Occasionally, from the accumulation of urine, the over-distended bladder is burst. In all cases after death the bladder is found im- mensely enlarged, and resembling in size the bladder of an ox instead of a sheep. The urethra is also greatly increased in size, its walls thickened, and its channel usually throughout a large portion of its extent blocked up by gritty, slimy matter. The kidneys are soft, flabby, and friable, and in most cases the ure- ters are enlarged, and the urine, unable to find its way into the already overfilled bladder, permeates the cellular textures about the loins, and thus renders the mutton of sheep slaughtered in the latter periods of the disease unsavory, unfit for keeping, and some- times scarcely fit for eating.

From what has been stated, it will be evident that the cure of these urinary disorders must depend upon clearing, if possible, the sediment out of the urethra. With rams in which the penis can be grasped and drawn out of the sheath, by careful manipulation the tube may often be cleared of its obstruction. Occasionally, however, the sediment is so gritty and lumpy, that even the most dexterous handling fails to displace it. Owing to the tortuosities of the urethra, the catheter, which is so useful in relieving such diseases in other animals, is of no service amongst sheep. When manipulation and hot fomentations fail, the only resource is to cut into the urethra at or behind the " worm,' and allow the escape of the deposit. In wethers fit for the butcher, as such cases usually are, it is seldom worth while to attempt treatment, and before the symptoms become aggravated, and the mutton injured, the sheep had better be slaughtered.

To prevent such disorders a change of diet should at once be adopted ; cake, corn, and other richly nitrogenous food should be given in lesser amount ; mangel should also be sparingly used ; if left cut for a few days before being given, it usually appears less injurious. The sheep had better be run upon some grass or seeds, where they will take sufficient exercise, and have a good supply of water. Any animal in the earlier stages of the disorder, which it is not thought desirable to slaughter, should have a dose of castor-oil and a drench daily of half. a drachm of carbonate of potash, which will help to neutralise the acidity of the urine.

When disposed to lie moping about, they should be frequently put on their legs and driven about. When indisposed to feed or drink, drench them with thin gruel, barley water, or even with plain water in which a little carbouate of potash is dissolved. From time to time try, by manipulation, to get away a little of the obstructing sediment.

BRAXY, OR SICKNESS.

This very fatal malady is confined to hoggs ; that is, lambs during the first winter. In the Highland sheep farming districts in Scotland as many as from 40 to over 60 per cent. of the hoggs have perished in consequence of it in the course of a few months, and the average of deaths in many districts has been reckoned at 20 to 25 per cent. It is a disease of a typhoid form, and runs its course so rapidly that numbers of hoggs which appear healthy at night will be found dead in the morning.

Causes.—The disease usually attacks hoggs in good condition, and follows pasturing on hard, rank, indigestible grass, from weaning and afterwards; also from exposure while on such pasture to sudden and violent changes in the weather; and from keeping the hoggs on one run of grass until it becomes foul.

Symptoms.—The first usually observed is that the hogg affected lies down frequently when the rest are grazing. It never eats, and there is a restlessness and occasional change of position, with a dull, sick look. The bowels are constipated; the head hangs down; the eye becomes glazed, and there is sometimes a rapid movement of the hind feet and a crunching noise with the teeth, indicative of pain. The paunch begins to swell, and the back rises; then the animal crawls away from the rest and lies down, but often rises and changes its position. This it continues to do until moving becomes so painful that it ceases to rise, and lays its head on the ground, a sad picture of patient, silent suffering. It allows itself to be caught, and makes but faint efforts to get free, then drops down, and makes no further attempt to get away. The swelling increases, and so does the agony of the animal, which is extreme, till death comes to its relief. There is frequently froth, mixed with blood, in the mouth and nostrils. On examination after death, the first thing noticed is the intolerable stench which arises from the viscera and carcass. The manyplus is found full of hard, compressed, undigested food, same as in "Fardel-bound" in cattle; the inner coating of the paunch comes off in large sheets or patches; the liver in whole or in part, but always in the neighbourhood of the gall-bladder, is in a state of decomposition; and the cavity of the body is often full of water tinged with blood.

Treatment.—When taken at an early stage, a strong dose of

Epsom salts, with some ground ginger in warm gruel, has been found useful. A correspondent of the *Irish Farmers' Gazette*, who states that he has succeeded in restoring several sheep attacked with braxy to perfect health, places his dependence chiefly on linseed oil given several times a day. He recommends 2 oz. linseed oil and 3 grains powdered opium in a little linseed tea. Repeat the opium the following day, with a scruple of ginger, two scruples of gentian, and if the bowels continue costive, the oil may be again administered. This treatment is in accordance with that recommended by Mr. Spooner. Bleeding in the early stage has also been useful in some cases.

Preventive Treatment.—More depends on this than treatment after the disease has appeared. The chief preventives are wholesome, succulent food and shelter. Turnips and oil-cake have been found to act in a great measure as preventives.

COLIC.

Lambs occasionally suffer from Spasmodic Colic.

Symptoms.—Severe pain, in paroxysms.

Treatment.—Administer slowly and carefully—

Tincture of opium	... ...	1 drachm.
Powdered ginger	... ...	1 ,,
Epsom salts ...	... ...	4 ,,

To be dissolved in warm gruel, and repeated, without the salts, if required.

CONCRETIONS IN THE STOMACH.

Sheep sometimes swallow considerable quantities of earth, apparently naturally impelled to do so, owing to acidity in the stomach ; and if the quantity swallowed is unusually large, irritation of stomach and intestines will be produced. Numerous balls of earth and wool are frequently found in lambs.

Treatment.—Epsom salts as a direct purgative ; and put down rock salt, and also chalk, in the pastures. Half a drachm to one drachm of carbonate of soda may be given where there is acidity of the stomach; the dose for lambs being proportionally less. Dissolve the carbonate of soda in a wineglassful of water.

DIARRHŒA.

This disease, which is most prevalent among hoggs and young sheep, is frequently brought on from want of attention to changing the pasture. Clean pasture is essential to the health of young sheep, and scouring, as this disease is sometimes called, is often prevented or allayed by merely changing them from a field in which they have been going for some time to fresh, although not luxuriant, pasture. Free access to rock salt is also of advantage. The following cordial medicine may be given :—

Powdered catechu	... ...	4 drachms.
„ prepared chalk ..	...	1 ounce.
„ ginger	... ...	2 drachms.
„ opium	... ...	½ drachm.

Mix with half a pint of peppermint water, and give twice a day, two or three tablespoonfuls to a sheep and half the quantity for a lamb; or

Powdered opium	... ...	2 grains.
„ gentian	... ...	1 drachm.
„ ginger	... ...	1 „

To be mixed in an infusion of linseed.

Diarrhœa assumes different forms in lambs. The " Green Skit" may be treated as above, a dose of 2 drachms of Epsom salts being given previous to the cordial medicine above mentioned.

" White Skit" is a more dangerous disease, and arises from the coagulation of the milk in the fourth stomach, and it is the whey passing off which gives the white appearance to the dung.

Symptoms.—A white appearance of the dung; dulness; heaving of the flanks; hardness and distention of the belly, and sometimes costiveness.

Treatment.—Give half an ounce of magnesia dissolved in a considerable quantity of water, or a quarter of an ounce of hartshorn, also in water; follow with a dose of Epsom salts, and then the cordial medicine as above. Carbonate of soda dissolved in a wineglassful of water is a useful remedy in the case of young lambs.

Sometimes diarrhœa show itself in young lambs when on their mothers; the dung being very dark in colour, and sometimes tinged with blood. This is a fatal form of the disease, and as it appears to originate in the ewe's milk, the food of the ewes should at once be changed. More dry food should be given, and a little oil-cake and malt, with some salt mixed with it, should form part of the food.

HOOVE.

This term is applied to distension of the rumen or paunch with gas, owing to the fermentation of the food.

Causes.—Sudden turning of sheep into a field of red clover or other very succulent food, especially at night or early in the morning, when dew or hoar frost is on the herbage. We have known sheep affected with it when first turned on rape, and on swede turnips having luxuriant tops.

Symptoms.—Swelling of paunch, same as in cattle.

Treatment.—Use the probang if at hand, or give 2 to 4 drachms hartshorn in half a pint of warm water; or a drachm or more of chloride of lime dissolved in water; or, failing those, a dessert spoonful of salt dissolved in water. It may be necessary to stab the animal in the flank; if this is done with a pen-knife, insert a quill to allow the escape of the gas. Give the following draught : —

Epsom salts	...	...	.. 2 ounces.
Gentian	...	...	... 1 drachm.
Ginger	..	...	... 2 drachms.
Chloride of lime	...	...	1 scruple.

K

Dissolve in warm water or gruel. It has been recommended as a preventive to sprinkle luxuriant herbage with salt previous to sheep being turned on it.

POISONING.

Sheep, as well as cattle, have often died in consequence of eating plants of a poisonous nature, such as hemlock, yew, &c. It is also dangerous to allow sheep to graze on pasture recently top-dressed with soot.

Symptoms.—Paralysis and constipation.

Treatment.—Sheep poisoned by eating hemlock have been saved by bleeding, and a solution of Epsom salts, acidulated with sulphuric acid. Large doses of linseed oil are also useful in cases of poisoning.

RED WATER.

This is a different disease from that in cattle known by the same name : it chiefly affects lambs when feeding on turnips, or other green food, particularly during hoar frost. It consists in effusion of red water in the abdomen, outside the bowels, and is a very dangerous and rapid disease.

Symptoms.—Loss of appetite and rumination ; dulness ; costiveness ; occasional giddiness.

Treatment.—If the animal is fit for the butcher, kill it as soon as it is observed to be affected. Remove the sheep from the turnip field, and allow them to remain there only a portion of the day. Mr. Spooner mentions that it is stated on respectable authority that to give young sheep or hoggets a table-spoonful of common tar every fortnight has been found a successful preventive. If medicine is to be given on the appearance of the disease, use the following :—

Epsom salts ...	...	...	1 pound.
Powdered ginger	...	..	1 ounce.
Powdered gentian	...	...	1 ounce.
Powdered opium	...	...	$\frac{1}{2}$ drachm.

Mix. This is sufficient for 8 or 10 sheep, or double the number of lambs. Dissolve in warm water or gruel.

ROT.

Causes.—Feeding on wet, marshy pastures, especially such as have a cold, stiff subsoil; wet seasons; overstocking.

Symptoms.—In rot there are always insects called flukes present in the ducts or passages which convey the bile; the muscles, after death, are found soft and flabby; the kidneys pale; the belly filled with water; the heart enlarged and softened; the lungs filled with tubercles; the liver pale, or curiously spotted; parts of it hard, and others ulcerated, and it is easily broken down with the slightest pressure. The external symptoms during life are: —The eye becomes injected, but pale, the small veins at the corner of the eye are filled with a yellow fluid, and as the disease becomes confirmed the yellow tinge spreads; the muzzle and tongue are stained; the animal is dull; loses flesh rapidly; the breath stinks; bowels are variable, being sometimes costive and at other times excessively loose; the skin becomes spotted with yellow or black, and is loose and flabby, giving forth a peculiar crackling sound when pressed upon; the wool comes easily off; the belly begins to enlarge; the animal becomes weak, and wastes to a skeleton, and there is a swelling or "poke" in the upper part of the throat.

Treatment.—Once the rot has fairly set in it is incurable. If taken in the early stage, removal to sound, dry pasture, with free access to rock salt, will have a good effect. The following in the form of a powder may also be given daily:—Common salt, $1\frac{1}{2}$ drachm; sulphate of iron, $\frac{1}{2}$ drachm. Feed liberally on cake and corn, and sell the animals to the butcher when fit for use.

Prevent the disease by thorough draining and moderate stocking.

WORMS.

Lambs are sometimes affected with worms.

Symptoms.—Depraved appetite.

Treatment.—Administer for the purpose of allaying irritation :—

Linseed oil	2 ounces.
Powdered opium	3 grains.

Mix.

After the irritation is diminished give,

Linseed oil	2 ounces.
Oil of turpentine	4 drachms.

To be well shaken. Change the food.

The following has also been recommended for worms in lambs : —Common salt, four parts ; powdered areca nut, two parts ; and of turmeric, one part. The powder may be mixed with bruised oats, meal, or other food, so as to allow half an ounce to each animal in the day.

CHAPTER IV.

DISEASES OF THE SKIN AND EXTREMITIES.

SCAB.

This very troublesome disease is akin to the itch in man, and mange in horses.

Causes.—The primary cause is the presence of minute insects, which burrow under the skin, and produce much irritation. The disease is readily propagated by contagion, derived either from direct contact with diseased animals, or when healthy sheep are allowed even to touch gate posts, trees, &c., upon which scabbed sheep have previously rubbed.

Symptoms.—Excessive irritation, causing the animal to rub against any hard object. The wool becomes broken in consequence of this rubbing. Pimples are formed on the skin, which becomes rough. In a short time, if the malady is not checked, the animal assumes a miserable appearance. In the early stage, even when the wool is not broken, a sure sign of scab may be observed in the following manner:—When the sheep is standing quietly in a pen rub the skin on the back gently with the fingers, and if the animal appears to like the operation, and turns its head round towards the operator, champing its teeth at same time, it is affected with scab.

Treatment.—There are various remedies for the cure of scab, and also preventives in the shape of "dips" and "pouring mixtures." These are all more or less efficacious; but there are serious objections to the use of some, which should be taken into consideration. Arsenical preparations have been fatal to sheep and other animals, especially when the sheep were exposed to rain soon after being dipped, causing the dip to be washed out of the fleeces upon the grass. Corrosive sublimate or bichloride of mercury is also largely used in certain sheep dips; and it appears to have the effect of causing sheep which have been frequently dipped in such baths to lose their teeth at an early age. Carbolic acid is an excellent cure for scab, but the stain it gives the wool is so permanent that it is almost impossible to wash it out; and in this way wool dressed with carbolic acid mixtures becomes much deteriorated in value to the manufacturer. Tobacco juice is now allowed to be manufactured free of duty, for the purpose of using it as the basis of sheep dip; but, in accordance with the law, it is required that the manufacturer shall mix a certain proportion of arsenic with the juice. The quantity of arsenic used for this purpose, however, is comparatively small, and a benefit rather than otherwise. Tobacco juice should not be mixed with carbolic acid. There are special dips prepared by certain manufacturers which neither discolour the wool nor are injurious to the health of the animals, and at the same time are perfectly efficacious for the purpose for which they are intended.

As scab is a very common disease, we give a variety of receipts, as follows :—

I.

Tobacco juice	...	...	...	5 lbs.
Powdered hellebore	...	...	...	¼ do.
Black soap	...	...	...	1 do.

Boil the soap and hellebore together in water for twenty minutes; when cool, add the tobacco juice, and make up to five gallons with water. This will be sufficient for "pouring" twenty sheep; and when a bath for dipping the same number of sheep is required, prepare the materials as above, and add water to make up to twenty gallons. Give each sheep a full minute in the bath.

II.

Arsenious acid	...	...	...	2 lbs.
Carbonate of potash		...	...	2 do.
Boiling water	...	...	...	60 gallons.

Mix, and boil for half an hour.

III.

Arsenious acid	...	...	...	2 lbs.
Sulphate of iron	...	...	...	20 do.
Water	...	...	...	60 gallons.

Mix, and boil until the fluid is reduced to a third, and then add as much water as has been lost by evaporation.

IV.

Arsenious acid	...	...	... 2 lbs.
Sulphate of zinc	...	...	... 10 do.
Water	...	...	... 60 gallons.

Prepare as the foregoing receipt.

V.

Powdered arsenious acid	...	... 4 lbs.
Powdered sulphate of iron	...	... 40 do.
Powdered block oxide of iron		... 1½ do.
Powdered gentian root	...	... 12 ounces.

To be well rubbed down and pressed in a well-closed jar. When used, it is mixed with ten times its weight of water, and boiled for ten minutes. The amount here mentioned is given as enough to dip 200 sheep.

VI.

The following composition has long been in repute in Scotland. It is most effectual as a cure for scab, but it discolours the wool. For one sheep, one wineglassful of spirit of tar, mixed with one bottle of water; for twelve sheep, one bottle of the spirit of tar, mixed with twelve times the quantity of water. For a hundred sheep, take six gallons of water and six pounds of common washing soda; warm the water to the boiling pitch, and then add spirit of tar in proportion as above. Large sheep will require more; and add cold water to make up the quantity required. This mixture is applied by "pouring," which is conducted as follows:— The shepherd sheds or divides the wool, and a boy or girl with a small tin pot, shaped like a teapot, but much smaller at the outlet, measures out the exact quantity, and pours out the liquid slowly, following up the furrow behind the shepherd's hands. The sheds, or furrows, ought not to be more than one inch apart, and wherever the scab has got a hard crust, a bluut knife ought to be used to scarify or scratch off the scab. The shepherd ought also to have a quantity of the pure spirit of tar at hand. If the animal is much affected about the hind quarters, particularly about the tail or thighs, or even about the forearms, he may apply the pure spirit freely, and rub well in with the hand. If one dress-ing is not sufficient to cure the scab, a second rarely fails.

TICKS.

These are troublesome animals, which infest the skin of sheep. The ordinary sheep dips in use for scab will remove them. (See " Scab.")

The following passage occurs in the report of the Governors of the Royal Veterinary College for the year 1869 :—

" Some other novel cases occurred in lambs, in which death resulted from parasites existing on the skin. The parasites in ques-tion were those commonly known as ticks (iodes ricimus). These parasites abound in most countries, and are met with both on wild and domesticated animals, firmly attached to the skin, from which they draw blood as their food. Until now they have not been found on animals in Great Britain to any extent injurious to health, much less as causing death. In hot countries, however, and particularly in many parts of South America, these parasites attack animals in such vast numbers, that even oxen succumb to the irritating and exhaustive effects of their attacks. Specimens of the skin of lambs, thickly covered with these epizoa, were sent from Kent by a veterinary surgeon consulted on the case. In his communication he writes that 'they had attacked the sheep and lambs, both on uplands and marshes, and that one farmer found a large quantity of them on some colts which were at pas-ture near to the sheep.' There are few parasites more tenacious of life than ticks ; but experiments having shown that they can easily be destroyed by carbolic acid, it was recommended that a trial should be given to dipping the sheep and lambs in a diluted mixture of the acid. This proved most effective in the destruction of the ticks, and thereby prevented a further loss of lambs."

WILDFIRE

Is a variety of Erysipelas which usually appears in autumn or the beginning of winter.

Symptoms.—The skin of the breast and belly inflames and rises into blisters, containing a reddish fluid, which escapes and forms a dark scab. The animal sometimes fevers.

Treatment.—Give a dose of Epsom salts, and apply some carbolic acid ointment or camphor ointment to the scabs.

FLIES OR MAGGOTS IN SKIN AND HEAD.

Flies attack sheep in warm, close weather, and lay their eggs upon the skin, where they hatch, producing maggots. In the case of sheep affected with diarrhœa, the eggs will be deposited in the tail; also wherever there is any accidental wound. The head is frequently attacked.

Treatment.—When the maggots exist among the wool, it must be carefully clipped away, the part cleaned and dressed with common train oil, to which a little spirit of tar may be added. Carbolic acid, one part, and glycerine, four parts, will also be found a useful dressing. Where the head is affected, the same treatment may be followed, or the sores on the head may be dressed with the ointment of carbolic acid, and covered with a linen cap. Carbolic ointment or train oil may also be used as a preventive.

FOOT-ROT.

This is essentially a disease of rank and moist pastures. It is never found where sheep have a wide range of dry pasture on hard land. It is not contagious, but the cause which produces it in some sheep in a flock will also produce it in others subject to similar influences. It is most prevalent in summer and autumn, although cases are to be met with at other seasons.

Symptoms.—Lameness in one or more of the feet, generally the fore feet. As the disease advances, the affected sheep are obliged to crawl on their knees, and pine away to skin and bone. On putting the finger between the hoof, the foot is found to be unusually hot and generally a little swollen. The crust or horn along the outside of the foot and near the toe either overlaps the sole, or part of it is broken away and separated from the hoof, and particles of earth or filth are accumulated in the cavity. Ulcers are formed, which tend still further to separate the hoof, and a thin stinking matter is discharged, and by degrees the foot becomes a mass of ulcers and proud flesh. In some cases the disease breaks out in the inside of the toe, and progresses beneath the horn around the outside of the foot, and sometimes the heel is first affected. In lambs and hoggs it generally commences between the hoofs near the heel, the horn being comparatively sound. In these cases, when the animal is examined in the first stage of the disease, the skin presents a red and tender appearance, but as the disease advances the flesh becomes putrified, and a thick, adhesive matter accumulates. By degrees the malady extends round the heel beneath the horn till the whole foot is affected ; and if allowed to run its course for a length of time, a cure is seldom effected until the whole crust is taken off on the side of the foot affected.

Treatment.—Pare away all the loose parts of the crust or horn, taking care not to draw blood. Clean the foot well with a coarse cloth, and anoint the diseased part with butyr of antimony, putting on as much as the sore absorbs. Keep the sheep resting on its rump for a few minutes, till the application takes effect, which is shown by the proud flesh or ulcers assuming a colour like that of toasted cheese. Put the animal into a dry paddock where the grass is bare. Examine again in two days, and if the crust still appears to separate from the hoof, the knife must again be applied, followed by the butyr of antimony as before. Examine at several intervals, and repeat treatment, if necessary, until a cure is effected. The grand secret of success in treating foot-rot is, to take it in the early stage, as every remove from that point renders it more obstinate and difficult to overcome. When dressing foot-rot, a dry day ought always to be selected.

It has been found very useful in cases of foot-rot to lay down a quantity of hot lime in a pen, and allow the sheep to stand upon it for a short time each day ; or the lime may be laid down in a gateway, through which the sheep should be driven twice a day. When the lime becomes hardened from treading, rake it up so as to loosen it well.

Professor Simonds recommended the application of lunar caustic to the ulcers and proud flesh, following that up with a dressing of tincture of myrrh. Lord Berners stated at a meeting of the London Farmers' Club that he had found this mode of treating foot-rot very successful.

A mixture of one part of carbolic acid with four parts of glycerine may also be employed in cases of foot-rot. The ointment of carbolic acid, one part of the acid to six parts of lard, is another useful form of using this valuable application for sloughing and unhealthy wounds.

Hogg recommended the following mixture to be applied to the diseased parts in foot-rot :—Oil of turpentine, 2 ounces ; sulphuric acid, 2 drachms ; to which may be added olive oil, 1 ounce.

The following ointment for foot-rot was recommended by the editor of the *Agricultural Gazette* in " answers to correspondents:" —Plaster of Paris, powdered, 1 ounce ; sulphate of zinc, powdered, 1 ounce ; creosote, 1 scruple ; Stockholm tar, 4 ounces ; lard, 4 ounces ; to be made into an ointment, and applied occasionally to the feet and between the claws.

Mr. Spoouer recommends the following to be applied after paring away the loose horn:—

Tar ...	...	...	...	8 ounces.
Lard ...	...	...	...	4 ,,

To be melted together ; then add slowly and carefully—

Oil of turpentine ...	...	...	$\frac{1}{2}$ oz.
Sulphuric acid, by measurement	...	$\frac{1}{2}$ oz.

The same authority also states that equal parts of hydrochloric acid and tincture of myrrh and aloes has been used with success.

CHAPTER V.

SPECIAL DISEASES.

FOOT AND MOUTH DISEASE.

Sheep are liable to this disease, which is fully described under the diseases affecting cattle.   Sheep, however, seldom suffer much from disease of the structures of the mouth in this malady, the principal affection being in the feet; vesicles occur between the claws, and sometimes on the outer side of the foot, from which point separation of the hoof commonly commences.   In severe attacks there is an affection of the mouth and tongue   We refer to the article in which this disease is treated with reference to cattle. Some of the milder applications recommended for foot-rot, such as carbolic ointment, will be useful as dressings for the feet.

SMALL-POX.

This disease, called *variola ovina*, was unknown until British ports were opened to the free importation of foreign sheep and cattle. It is a formidable and fatal disease, and requires to be very strictly guarded against, which can only be effected by a most rigid inspection and quarantine of all sheep brought from the continent of Europe.

Symptoms.—A dull, moping appearance; dulness of the eyes; swelling of the eyelids, and reddish spots on the naked places. In a few days swellings resembling flea-bites appear on the skin, which in mild cases are moderately red and circumscribed; but in some cases are of a purple hue, and sometimes run into each other.

Treatment.—The disease is highly contagious and infectious, and treatment is of so little avail that it is useless experimenting in order to attempt a cure. Slaughter the animals, and bury them, in their skins, in a deep hole, putting plenty of quicklime on the carcasses.

ABORTION

This is far from uncommon in ewes, and is traceable to different causes, such as hurrying ewes heavy in lamb with dogs, either the shepherd's dog or strange animals. It is also supposed to be occasionally caused by an over liberal use of salt, although this is doubtful, as we have known ewes to be fed during winter on very heavily salted hay, without the slightest bad consequences. Abortion certainly follows the free use of turnips or mangels by ewes when heavy in lamb, instead of which they ought to have corn and cake, merely allowing a very limited supply of roots, say not more than a single turnip to each ewe per day until lambing has taken place, when the ewes may get as many turnips as they will eat. Wet weather in autumn and early part of winter is also conducive to abortion.

Treatment.—Usually treatment is unnecessary, beyond placing the ewe for a few days in a house or sheltered place, and nursing her. If there is any fever or costiveness, give an ounce or two of Epsom salts, in a little gruel; and should she continue weak and indisposed to eat, give a little gentian and ginger, or malt. In fact, a little malt may be given daily under almost any circumstances.

PART IV.

DISEASES OF SWINE.

CHAPTER I.

THE HEAD, NERVOUS SYSTEM, AND ORGANS OF RESPIRATION.

APOPLEXY.

Causes.—High-feeding, accompanied by want of exercise. This disease is usually sudden and fatal, and even when relieved, it frequently runs on to inflammation of the brain.

Symptoms.—Dulness; disinclination to move; heaviness of the head; an uncertain and staggering gait; wildness of the eyes, and blindness.

Treatment.—See "Inflammation of the Brain."

INFLAMMATION OF THE BRAIN.

Causes.—Heating or indigestible food; an over-feed of grains; new corn, &c.

Symptoms.—Dulness; redness of the eyes; disinclination to move; the animal runs wildly to and fro, and seems blind and unconscious where he is going.

Treatment.—Bleed from the palate if possible; make slits in the ears, which wash with warm water to encourage the bleeding. Give a dose of Epsom salts, say 2 ounces to a grown pig, with some ground ginger, or a dose of 2 to 4 ounces of castor oil, and 2 to 4 drachms of jalap, mixed. Give injections of soap and warm water, after which give repeated doses of 1 to 2 ounces of sulphur in milk or gruel. A blister may also be applied to the throat. Give nitre in the water the animal drinks.

EPILEPSY.

This disease is by no means uncommon in pigs.

Causes.—The causes are somewhat obscure. Youatt says that the immediate cause is generally some excitant or stimulant acting on a system predisposed by cerebral inflammation, or by intestinal irritation, arising from worms or other sources, to take on disease. In young, highly-bred pigs it is also believed to arise from over-feeding and indigestion.

Symptoms.—Constant grunting, restlessness, acceleration of breathing, pallor of the skin, and a staggering gait. The animal falls suddenly, lies for a few moments motionless; after which convulsions come on—at first gradually, but rapidly increasing in intensity; the neck is curved in every direction; the legs are alternately drawn up to the body, and then rapidly extended. The eyes protrude, and the balls roll about. The tongue is also protruded and fixed between the clenched jaws; the teeth grind together, and the animal foams at the mouth. When the animal gets over the fit and rises, he tries to hide himself, and looks terrified and wild for a time. If one animal in a sty is affected with epilepsy, the rest are also liable to be attacked.

Treatment.—It is only between the fits anything can be done. Apply at once a cooling lotion to the head; bleed, and give an active purgative medicine, as specified in "Inflammation of the Brain." A cooling lotion for the head may be composed of a pint of vinegar to two quarts of water, and one ounce of sal-ammoniac; or take equal parts of common salt, nitre, hydrochlorate of ammonia, and dissolve in a sufficient quantity of water. Apply these lotions with cloths dipped in them, and kept constantly wet.

A correspondent of the *Irish Farmers' Gazette* gives the following mode of treatment as one which he has found successful :—

" Bleed, purge, and put the patient in a box, without lid, by the fire ; tie a cloth round the neck, having a good pad at the nape ; let a girl keep this continually wet, night and day. She should do nothing else than to attend to her patient, holding him in his fits, to prevent his injuring himself. For the same reason the box should be a good tight fit ; if too large he will surely hurt himself. The addition of vinegar to the water is, of course, good ; but pure cold water *on the instant* is worth all the lotions in the world *with delay.* The warmth from the fire has the effect of keeping the blood in the extremities when forced there by the cold application to the head. It is not advisable to bleed sucking pigs, except as a *dernier ressort.* Twenty-four hours will see the patient able to dispense with a night attendant, and feed greedily, but ' short commons' must be the order for several days."

LOCKED JAW.

Causes.—Castration, followed by high feeding ; exposure and heating in travelling to market, &c.

Symptoms.—Spasmodic motions of the head and of one or more of the extremities ; grinding of the teeth ; rigidity of the jaws ; stiffness of the neck ; and an unnatural, upraised position of the head.

Treatment.—Bleeding ; warm baths ; purgatives ; the best

L

being 3 to 6 drops of croton oil in a little oil, as the most likely to be introduced ; and injections, such as the following :—

Linseed oil	...	...	1 pint.
Oil of turpentine	...	...	4 oz.
Croton oil	...	...	½ drachm.
Decoction of oats	...	...	2 pints.

Mix.   Rub well into the skin of the jaws and neck the following stimulating liniment :—

Oil of turpentine	...	...	1½ oz.
Tincture of cantharides	...	...	1½ „

Mix.   Locked jaw is often speedily fatal.

CATARRH OR COLD.

Symptoms.—A cough, and discharge from the nostrils.

Treatment.—In ordinary cases warmth and careful nursing will generally effect a cure.   In severe cases give the following medicine daily for several days :—

Antimonial powder	...	..	2 to 6 grains.
Nitre	...	...	10 to 30 „
Digitalis	...	...	1 to 2 „

In addition to this, rub a stimulating liniment on the brisket, such as the following :—

Soap liniment	...	...	2 oz.
Compound camphor liniment	...	...	2 „
Tincture of opium	...	...	½ „

Mix.

STRANGLES OR QUINSY

This malady is most common in fatting pigs, and being rapid in its progress, is generally very fatal.   It is also so infectious that any animal attacked with it should be at once removed from the rest.

Symptoms.—The glands under the throat swell; swallowing becomes difficult; respiration is impeded; hoarseness and debility then follow; the pulse becomes quick and unequal; the head partially paralysed; the neck swells, and rapidly goes on to gangrene; the tongue hangs from the mouth, and is covered with slaver.

Treatment.—In the early stage nitre dissolved in the water the animal drinks will, with attention to diet, care, and warmth, be sufficient to check it. The dose of nitre for each animal is from one to two drachms. If the disease goes on, bleed and purge, and insert setons in the swollen glands, or apply blisters and strong external stimulants. For such stimulants see " Catarrh."

INFLAMMATION OF THE LUNGS.

This is a prevalent and frequently fatal malady. It sometimes assumes the form of Pleurisy or Bronchitis, and runs through the whole piggery.

Causes.—Atmospheric influences; unwholesome food; damp, ill-ventilated sties.

Treatment.—Bleed from the palate; open the bowels moderately, without purging, by giving Epsom salts and sulphur in a dose of from two to four drachms of each. The following medicine may then be given:—

Calomel	..	...	... 1 to 3 grains.
Tartarized antimony		...	... 1 to 3 „
Nitre...	...	...	... 5 to 20 „

After one or two doses the calomel may be omitted. Blisters may also be applied to the chest. Keep the pigs warm and clean.

CHAPTER II.

THE INTERNAL ORGANS.

~~~~~~~~~~~~~~~~~~

COLIC.

Causes.—Unwholesome food; cold, or wet, filthy sties.

Symptoms.—Restlessness; sudden and extreme pain, evinced by cries, and rolling on the ground.

Treatment.—A dose of tincture of opium and spirit of nitrous ether, from one drachm to eight of the former, and double this quantity of the latter, according to the size of the animal, given in a few ounces of warm water. Or a dose of castor oil, proportionate to the size of the animal, with a little ginger in it, to be given in warm milk; repeat if necessary. Give warmth, comfort, and wholesome food.

INFLAMMATION OF THE BOWELS.

Cause.—Usually unwholesome food.

Symptoms.—Considerable pain, as in colic; constipation, great fever, and loss of appetite; is a very dangerous disease, even in its mildest form.

Treatment.—Copious bleeding from the vein on the inside of the forearm. Give two to four ounces of castor oil, and half doses every two hours until the bowels are opened. Give also injections of linseed oil, and warm baths; the latter more especially in the case of young pigs. Restrict the diet to the simplest and lightest food.

DIARRHŒA.

This disease is not uncommon, especially in young sucking pigs, in which case it arises from the sow having more milk than the young ones can take. Cleanliness and regularity in feeding tend to prevent this disease.

Treatment.—

Prepared chalk	...	...	..	1 oz.
Powdered catechu	...		. .	$\frac{1}{2}$,,
,, ginger	...	...	...	2 drachms.
,, opium	...		...	$\frac{1}{2}$,,

To be mixed and dissolved in half a pint of peppermint water. A teaspoonful twice a day will be enough for sucking pigs, and larger animals from half an ounce to an ounce of this mixture twice a day. Pay strict attention to the diet, which should consist only of dry farinaceous food. If the dung is slimy, give a dose of Epsom salts previous to administering the above cordial.

INFLAMMATION OF THE BLADDER OR OF THE KIDNEYS.

These diseases are not common ; at same time, inflammation of the bladder is sometimes produced by constipation, which, causing pressure on the neck of the bladder, produces inflammation.

*Treatment.—*Copious bleeding and doses of two to four ounces of castor or linseed oil, and in addition 1 to 3 drachms of laudanum to allay irritation. Warm baths are also useful.

SPLENITIS.

*Symptoms.—*Restlessness and debility ; swine suffering under it shun their companions, and bury themselves in the litter ; loss of appetite ; excessive thirst ; short breathing ; cough ; vomiting ; foaming at the mouth ; grinding of the teeth ; the groin is wrinkled, and of a pale brownish hue, and the skin of the throat, chest, and belly, which latter is hard and tucked up, is tinged with black.

Treatment.—Copious bleeding and purging, and the application of cold lotions of vinegar and water to the parts in the neighbourhood of the spleen, and of a cold shower bath, which may be given by means of a watering pot. The disease is usually very fatal.

PROTRUSION OF THE RECTUM.

Cause.—Chiefly obstruction of the intestines.

Treatment.—As this disease is frequently attended with rupture of some of the intestines, the animal should be slaughtered. If it arises simply from obstruction, wash the parts ; carefully return the rectum, pushing it up some little distance. Pass some strong thread, doubled, several times through the anus, and tie it with a knot ; no solid food should be given for some days ; only a little milk.

WORMS.

Symptoms.—The animal eats voraciously, but does not improve in condition ; coughs ; runs restlessly about, uttering squeaks of pain ; becomes savage, and snaps at its companions. The dung is hard and high coloured ; the eyes sunken ; and along with general debility, the animal has frequent attacks resembling colic.

Treatment.—Strong purgatives and doses of turpentine, 2 to 4 fluid drachms for a dose. Youatt says, "Nor must it be supposed that because no worms are seen to come away from the animal the treatment may be discontinued, or that there are none ; hundreds of them die in the intestines, and there become digested and decomposed."

CHAPTER III.

THE SKIN.

~~~~~~~~~~~~~~~~~

ERUPTIVE DISEASE OF THE SKIN.

Cause.—High living.

Treatment.—A dose of Epsom salts; cooling drinks—that is, one or two drachms of nitre and some sulphur in the water the animal drinks; lower diet, milk, whey, &c. Apply the following cooling lotion to the skin :—

Muriate of ammonia	...	...	4 drachms.
Acetic acid	...	...	... 1 oz.
Cold water	...	...	... 1 pint.

ERYSIPELAS.

This sometimes attacks highly-bred and highly-fed pigs.

Cause.—Apparently sudden transitions of temperature.

Symptoms.—Disinclination to eat; skin very hot; and large spots of a livid hue break out.

Treatment.—Warmth and quiet. Give cooling drinks, such as one or two drachms of nitre and a little sulphur dissolved in the water.

LEPROSY.

Causes.—Bad water; exposure to the inclemency of the weather; insufficient food, and damp, marshy localities.

Symptoms.—This is a very insidious disease, and the early symptoms often pass unobserved. It consists in the formation of vesicles or whitish granulations in all parts of the cellular tissue, and these vesicles contain a species of worm, somewhat similar to that found in the brain of sheep. These vesicles chiefly appear in the thigh or ham, the shoulders, around the jaws, along the neck and belly, and underneath and around the root of the tongue.

The general symptoms are :—Marked stupidity or obstinacy in the animal; a state of langour and general debility; a thickening of the skin; a tendency in the hair to fall off. The animal is often exceedingly voracious. As the disease increases in intensity, the animal shows great weakness, the breathing is slow, and a stinking smell is emitted; the hinder parts become paralysed; the skin comes off in large patches; the tongue is swollen and dark-coloured; the teeth are ground convulsively; the body swells; feeble cries of pain are uttered, and the animal dies.

Treatment.—Give one to two ounces of sulphur and antimony mixed; and repeated small doses of Epsom salts. Apply externally to the ulcerated parts an ointment composed of

Iodide of sulphur	... ...	1 drachm.
Glycerine ...	... ...	6 oz.

Attention to be paid to cleanliness and diet; only cooling, wholesome food should be given.

LICE.

Cause.—Filth.

Treatment.—First wash the skin thoroughly in all parts with warm water and black soap, and then rub in tobacco water or train oil, or diluted carbolic acid, not omitting any part. Repeat if necessary. Give a little sulphur and salt internally.

MANGE.

Cause.—Minute insects, same as in sheep (scab) and human beings (itch).

Symptoms.—Pustules or spots under the armpits and interior of the thighs; these gradually run together and form large scales, which the animal rubs into large, blotchy sores.

Treatment.—Rub well into the skin tobacco water, sulphur ointment, or carbolic acid ointment with which two parts of sulphur has been incorporated.

MEASLES.

Symptoms.—The seat of this disease is under the skin. It consists of a number of small watery pustules between the fat and the skin. Externally there are reddish patches, somewhat raised above the surface of the skin, on the groin, the armpits, and the inside of the thighs at first, and subsequently on other parts of the body. The other symptoms are heat of the skin, or fever; cough; discharge from the nostrils; loss of appetite; swelling of the eyelids; feebleness of the hinder extremities; blackish pustules under the tongue; and eventually the skin comes off in patches.

Treatment.—Measles are seldom fatal, and usually yield to simple, cooling treatment, such as doses of Epsom salts, 2 to 6 ounces, and one or two drachms of nitre, with cooling diet, and attention to temperature and ventilation.

CHAPTER IV.

SPECIAL DISEASES.

~~~~~~~~~~~~~~~~~~~

FOOT AND MOUTH DISEASE.

This malady attacks swine as well as other animals.

Symptoms.—Lameness in the feet, from soreness between the claws, and inflammation of the substance connecting the bone with the hoof, so much so that matter is formed, and the hoof is cast. Pigs are generally affected in the feet and muzzle, and sows suffer occasionally from the formation of vesicles in the udder.

Treatment.—See this disease under "Diseases of Cattle."

The following powder may be sprinkled on ulcers in this disease in the case of all kinds of stock :—

Powdered chalk	...	...	4 ozs.
„ charcoal	...	...	1 „
„ alum	...	...	$\frac{1}{2}$ „
Sulphate of zinc	...	...	$\frac{1}{2}$ „

Mix.

The following cleansing mixture may be applied as a dressing between the claws, and to the feet :—

Powdered alum	...	...	... 6 ozs.
Sulphate of copper	...	...	... 2 drs.
Oxymel	...	...	... $\frac{1}{2}$ lb.

Mix.

In his report on "Continental Farming and Peasantry," Mr. Howard, M.P., gives the following *recipe* for the foot-and-month disease :—"Take honey, 1 lb. ; muriatic acid, $1\frac{1}{4}$ oz. ; mix them well with a wooden spoon in an earthenware vessel; apply with a wooden spatula about a dessert spoonful to the tongue three times a day, leaving the animal to distribute it over the inside of the mouth by the champing motion which is sure to follow its application. For the feet, take aloes, ½ oz.; rectified spirit, ½ pint; alum, ½ oz.; dissolve in 1 pint of water; mix and apply a little twice a day between the claws." This may be used also in cases of foot-and-mouth disease occurring in cattle and sheep.

PALSY, OR PARALYSIS.

Causes.—Low, marshy situations; bad or damaged food; also highly stimulating food.

Symptoms.—The hinder parts become paralysed; the animal is unable to rise, or falls when attempting to walk.

Treatment.—Wholesome food, clean straw, a clean, dry, and well-ventilated sty, moderate exercise, and gentle doses of Epsom salts, or of common salt with 3 scruples to a drachm and a half of gentian or ginger powdered in addition.

RHEUMATISM.

Cause.—Cold and damp sties.

Treatment.—Colchicum in doses of from 2 to 5 grains, repeated daily for three or four days. Give also a dose, 2 to 6 ounces, of Epsom salts, with some powdered ginger. Remove to a warm, dry sty, with plenty of litter, and keep the animal comfortable.

Mr. Spooner states that there is an affection of the joints in pigs, resembling rheumatism, which is caused by the food or water being impregnated with lead. Some kinds of water act on lead pipes much more freely than others.

The remedy consists in the removal of the cause. Give also a dose of Epsom salts; say 2 to 6 ounces.

RED SOLDIER.

This fatal malady has only become recently prevalent. The

following extract from a report made by Professor Ferguson to the Irish government in 1867 describes its nature fully :—

" For the last three years a most malignant and fatal disease has made its appearance each summer as an epizootic among swine in Ireland. The principal characters of the malady are the suddenness of the attack, without any premonitory symptoms, its rapid course, and great mortality. It frequently happens that the animal that had been observed feeding well and in apparently good health half an hour before, is found either dead or dying. The skin, especially on the abdomen, and other parts where it is thin, becomes of either a purple or red colour, sometimes in more prolonged cases so red all over that the disease is called ' the soldier,' from the colour of the skin of animals so affected with it resembling that of the uniform of the British army. Pigs affected with this disease are generally known as either ' soldiers' or ' red coats.' The disease is both infectious and contagious, although there is ample evidence of its having made its appearance in piggeries the porcine inhabitants of which had not for many months previously been exposed to any of the usually recognised infection carriers, excepting the ordinary atmosphere, which it would appear can convey the infectious principles of some diseases great distances, as it can some easily recognised powerful odours. Some of the outbreaks of this disease have made their first appearance among pigs in some particularly well isolated positions, being separated by a distance of miles from any other animals of the same species. It must be ceded that however infectious and contagious may be the disease in question, it is capable of being generated idiopathically with such influence. But having once made its appearance even in but a single case, it rapidly spreads through a very large percentage, but not all, of the pigs exposed to its influence. The system of separation of the healthy from the diseased animals, the isolation in different premises, have been found the most effectual means of arresting the progress and spread of the disease. The malady partakes markedly of the characters

of malignant scarlatina and typhus of the human subject, not alone
in its symptoms during life, but also in the appearance presented
after death by the different organs and tissues, excepting that in
this pig disease the fatal termination occurs as quickly as from
half an hour to forty-eight hours after the first apparent accession
of the malady, and that apoplectic congestions and effusions are
more frequent in this pig malady than in either of the above
human diseases. In the suddenly fatal cases, on *post mortem*
examination there are almost invariably found apoplectic effusions
of and within the brain and spinal cord. Similar effusions, in
almost all quickly fatal cases, are found in the substance of the
muscles, giving an irregularly bruised appearance to the flesh.
Even the muscular structure of the heart is frequently found thus
altered in its appearance. In many cases a pustular eruption very
much resembling that of foot-and-mouth distemper appears about
the feet, affecting the animal so painfully that it cannot stand ;
the bowels are generally exceedingly costive and difficult to move
by medicine, but in some cases there is diarrhœa in the early
stages : such cases seldom die suddenly, but when opened after
death the air passages of the lungs are found filled with a mucus,
which during life had caused severe coughing. In the protracted
cases, the meat is frequently found of its natural colour, quite
unlike the dark and blotchy appearance observed in the suddenly
fatal cases. Seventy-five per cent. is below the average mortality
of this disease. Unlike measles, it is found to attack with the
greatest frequency and severity pigs that are kept together in
great numbers, particularly when confined in sties, after being
collected together from different parts for fattening. It seldom
makes its appearance idiopathically among pigs that are allowed
to go at large. If it breaks out in a piggery, and that the season
is favourable, and the fences convenient, turning the sound
animals at large into either fields or plantations considerably diminishes the number of pigs attacked."

The flesh of pigs which die of this disease, or which are

slaughtered after becoming infected, is unfit for use as human food.

" Red soldier" is generally considered incurable, but a correspondent of the *Irish Farmers' Gazette* stated that a seton made of a strong fibrous part of the black hellebore plant acts as a preventive, and, he says, even a cure. Small quantities of sulphur occasionally in the food, and sulphite of soda—2 to 6 drachms—on the intermediate days, have been recommended as preventives.

Other correspondents of the *Farmers' Gazette* have recommended as preventives the frequent use of the cold bath ; cleanliness ; the administration of one to two tablespoonfuls of flowers of sulphur in the food, twice a week ; and also the following mixture, as a remedy :—1 teaspoonful of solution of chloride of lime, in a wineglass of water, given internally, and 4 teaspoonfuls of the same, in 4 wineglasses of water, well rubbed along the back and loins. To be given and applied so soon as the animal is observed to refuse its food. With reference to the cold water bath as a remedy, a correspondent of the *Farmers' Gazette* (August 27th, 1870) says :—" Swim them in cold water for ten minutes ; then cover them well in dry straw or chaff [or horse rugs], when they will sweat profusely." For different reasons we consider this as being well worth trying ; also the regular use of the cold bath, except in frosty weather.

SCROFULA.

Pigs which are too much bred in-and-in are most liable to scrofula. Tubercles are formed in the lungs and other internal organs, and the animal so affected dwindles away and dies.

Prevention is the only means of warding off this malady ; the means being obvious. Curative treatment is useless.

PART V.

DISEASES OF DOGS.

CHAPTER I.

THE HEAD, EYES, AND MOUTH.

CANKER IN THE EAR.

Causes.—Injudicious feeding on over-stimulating diet, such as flesh ; want of exercise ; damp kennels.

Symptoms.—The dog constantly shakes and scratches the ear ; a blackened discharge within the ear, which emits a smell like that of decayed cheese, and a crackling sensation is imparted to the fingers when the base of the ear, below the flap, is manipulated.

Treatment.—Use the following wash :—Goulard's extract and distilled water, of both equal parts. Two persons are required to apply the lotion. Youatt says—"The surgeon must hold the muzzle of the dog with one hand, and have the root of the ear in the hollow of the other, and between the first finger and the thumb. The assistant must then pour the liquid into the ear: half a teaspoonful will usually be sufficient. The surgeon, without quitting the dog, will then close the ear, and mould it gently until the liquid has insinuated itself as deeply as possible into the passages of the ear." After one ear is done, let it be covered closely with the flap, and the other side of the head turned upward, without releasing the dog.

As a rule, medicine need not be given, but keep the animal solely on vegetable diet. If it is necessary to give medicine, let it be a tea or dessert spoonful to an ounce, according to size, of castor oil or olive oil; but strict adherence to vegetable diet is an essential point in this, and in all diseases of dogs, as well as in their general management.

If large abscesses containing watery matter form inside the flaps of the ear, the animal must first be muzzled by binding a piece of tape thrice round the mouth, and connecting it by a piece passing over the forehead to a collar of tape, &c., round the neck. Place the dog between the knees, turn up the ear, and with a small lancet make an opening in what is then the upper part of the abscess. Introduce into the opening a straight probed bistnary by pressure, divide the sac or bag containing the fluid matter, which then escapes, and instant relief is given. Fill the opening for a day or two with lint soaked in the wash above mentioned, after which it is only necessary to keep the wound clean.

CATARACT.

Causes.—Mostly, but not always, old age; small-bred, high-fed dogs most subject to it.

Symptoms.—Total blindness; the crystalline lens becomes opaque.

Treatment.—Mayhew, who has devoted more attention than any other veterinarian to the diseases of dogs, says, with reference to cataract, that medicine appears to do injury rather than to produce benefit; and recommends cleanliness in the bed, whole-some food, a total abstinence from flesh, the daily use of the cold bath, and the constant use of a penetrative hair brush to the skin afterwards, combined with exercise.

GUTTA SERENA.

Causes.—Disease of the brain ; blows on the head ; also a temporary affection accompanying fits, or weakness, caused by keeping the dog, too long in the warm bath, or by over bleeding, &c.

Symptoms.—Blindness ; the eye remains perfectly clear, but the iris remains permanently fixed ; sudden light produces no effect on the eye; in the latter stage the eye changes colour.

Treatment.—If medicine is necessary, give a little castor or olive oil, and attend to general directions given in the case of cataract.

Blowing in pounded lump sugar and other gritty substances, as recommended by some, should be avoided in this and all affections of the eye. It is downright cruelty.

OPHTHALMIA.

Causes.—Dust, dirt, thorns, portions of leaves getting into the eyes ; result of some affection of the stomach.

Symptoms.—Constant closing of the eyelid, and a perpetual flowing of tears. On opening the lids, the inner lining is seen to be inflamed, and the membrane which covers the ball of the eye is of a white colour, and opaque.

Treatment.—If arising from dust, dirt, &c., remove the cause, which may be effected by a sponge or piece of linen dipped in warm milk and water. Take a large piece of lint, and double it several times ; saturate it with the following lotion, with which the lint which is laid on the eye must be kept wet :—Tincture of arnica, three drops ; tincture of opium, six drops ; camphor mixture, one ounce. When the dog can raise the eyelid, and when the flow of tears has become less, change the foregoing lotion for the following wash :—Nitrate of silver, one grain ; cam-

M

phor mixture or distilled water, one ounce. This is to be applied by means of a large-sized, long-haired, camel's hair brush; pour a little of the liquid into a saucer; saturate the brush with it; pull the lids gently asunder, and, having the eye exposed, draw the brush quickly over it. Beyond these remedies, Mayhew emphatically protests against any other treatment.

The above lotions are also useful when the eye has been scratched by cats, &c.

DEAFNESS

Causes.—May arise from cold, and if so, will disappear on the application of proper remedies for that affection.

Treatment.—If it does not arise from cold, and the cause appears obscure, wash the ears well with warm water and soft soap, and after well drying the ears, inject into them with a syringe sweet oil, in order to dissolve wax; but if there is no wax present, apply twice a week behind the ear, having first cut the hair close off the part, a blister of equal parts of oil of turpentine and solution of ammonia.

CANKER IN THE MOUTH.

Cause.—Decay of the teeth, causing inflammation of the bones of the lower jaw, and the gums.

Symptoms.—Redness and swelling; suppuration; an enlargement, which increases till a hard body is formed on the jaw, under the skin; powerful stench given off; proud flesh grows on the part; sinuses or pipes form in all directions, and bleeding is profuse.

Treatment.—Avoid caustics, which only make matters worse. Remove diseased stumps, and all the molars on the diseased side. Strengthen the constitution; for which purpose give cod liver oil twice a day, in doses of $\frac{1}{4}$ to $\frac{1}{2}$ an ounce, in milk. Wash the part with fomentations of a decoction of poppy heads, containing chloride of zinc in minute quantities. If the tumour remains stationary, the knife may be resorted to; but success is doubtful.

TEETHING.

Weakly pups will require attention, and when a tooth is loose, draw it at once. The temporary tusks of small dogs may also require to be drawn, if ton shed naturally when the permanent fangs should make their appearance. Moving the temporary tusks backward and forward with the finger will assist in loosening them. Attend to diet.

WORMING.

A needless and cruel operation, originating in ignorance.

FOULNESS OF THE MOUTH.

Causes.—Age; improper food; close breeding.

Symptoms.—The teeth grow black from an incrustation of tartar; the insides of the lips ulcerate; the gums bleed at the slightest touch; the breath is most offensive; the dog is afraid to eat, and even to drink; the throat is sore, and saliva dribbles from the mouth.

Treatment.—Extract all loose stumps, and lance the gums where necessary. Wash the mouth, &c., with a weak solution of chloride of zinc, a grain to an ounce of rose water, and flavour with a little oil of aniseed. In four days scale the teeth with instruments similar to those used by the dentist. Remove the incrustation by inserting the instrument between the substance and the gums, when it will come off in large flakes. Use great gentleness with the animal, who will be found to resist the operation. Take a small stencilling or poonah painting brush, and after saturating it with the above mentioned solution, clean the dog's teeth with it every morning, which will prevent fresh tartar accumulating. To strengthen the constitution, give $\frac{1}{4}$ to $\frac{1}{2}$ an ounce of cod liver oil once or twice daily, in milk.

PARALYSIS OF THE TONGUE.

Cause.—Unknown. Some dogs are pupped in this condition.

Symptoms.—The tongue hangs out of one side of the mouth, and becomes dry and hard, and the nose becomes hard and encrusted on that portion of its surface which the disabled tongue cannot reach.

Treatment.—Local treatment is useless. Let the diet be invigorating and wholesome, not stimulating in its nature, like flesh ; give an airy lodging ; plenty of exercise ; keep the digestion right, and give tone to the system. Treatment of this kind will alleviate and limit the evil ; but it cannot be removed even by dividing the muscles on the side on which the tongue is protruded.

CHAPTER II.

THE NERVOUS SYSTEM.

~~~~~~~~~~~~~~~~~~~~

FITS.

Causes.—Feeding much on flesh food ; occurs also from excitement, teething, pupping, and in connection with distemper.

Symptoms.—The dog stands as if stupefied ; emits a strange, loud, gutteral sound, and then falls on its side, still crying, but more feebly ; the dung and urine is passed involuntarily ; it will bite any one who attempts incautiously to lay hold of it ; the limbs, which at first are rigid, become violently moved ; the eye is protruded, and the animal foams at the mouth.   On recovering, the animal raises its head, stares about, and then runs off wildly, when it is usually mistaken for a mad dog, and doomed accordingly.

Treatment.—When the animal is on the ground, if from home when the fit comes on, put a handkerchief round the neck or through the collar, to prevent running away after the fit passes off.   Get him home as soon as possible, and at once administer the following injection :—Sulphuric ether, 1, 2, or 3 drachms ; laudanum, 2, 4, or 6 scruples ; cold spring water, $1\frac{1}{2}$, 3, or $4\frac{1}{2}$ ounces.   Leave the animal in silence for an hour, and then administer another injection of the same kind, repeating until the animal goes quietly to sleep, when it may be left to recover at its leisure.   Mr. Mayhew speaks very strongly of the beneficial effects of this treatment, in opposition to every other, and it has the merit of being very simple.   Avoid flesh as food, and keep the stomach in order ; wash seldom, and when done, with slightly tepid water.

RABIES, OR MADNESS

Cause.—Obscure; bite of a mad dog; exciting the nervous irritability of the animal to an undue exteut.

Symptoms.—In the early stages an alteration will be detected in the usual manner and bearing of a dog. He shows extreme restlessness, dulness, and stupidity, though he still recognises his master; he picks up straws, dirt, filth, and other unnatural substances; and one of the most suspicious symptoms is an uuusual desire for solitude, inducing him to creep away and hide in some dark, out-of-the-way place. When this is noticed, have him chained up at once, and call in the veterinarian to decide on the case. Instead of avoiding liquids, he will drink ravenously and lick up even his own urine. If allowed to remain loose, he starts off in a slouching manner on long journeys; his eyes are dull and retracted; his thirst increases, but as his throat has now become swollen, he is unable to drink, although he will plunge his head into the water. He snaps and flies at everything and pulls it to pieces. The noise he makes is incessant and peculiar. It begins with a bark, which is changed into a broken, interrupted howl. Paroxysms come on, and the last stage is reached.

Treatment.—Useless; shoot him at once.

When the blood has been ever so slightly drawn by the tooth of a dog, it is well, in order to prevent uneasiness, to touch the place at once with a stick of lunar caustic, moistened at the end, and then apply a bread and water poultice for five or six hours afterwards.

CHAPTER III.

THE THROAT AND RESPIRATORY ORGANS.

BRONCHOCELE.

Symptoms.—An enlargement which presses on the throat and jugular veins ; the animal is dull, the breathing laborious; sleeps much, and if a pup, in which the affection is most severe, the patient soon dies.

Treatment.—Generally useless in the cure of pups; but in that of an older animal, first shave off the hair from the enlargement, and then paint it over with the following tincture :—Iodide of potassium, one drachm ; spirits of wine, one ounce. This must be used freshly made up—that is, every three or four days at most. Administer iodine internally in the form of pills, from a quarter of a grain four times a day to a small pup, to two grains four times daily to a larger dog.

CHRONIC DISEASE OF THE WINDPIPE.

Symptoms.—A continued, confirmed cough, having a deep, sonorous sound. In bad cases every act of inspiration is followed by a kind of noise intermediate between a cough and a grunt, and it is sometimes accompanied by a species of roaring.

Treatment.—Let the animal have airy but comfortable lodgings ; no solid flesh to be allowed; but weak beef tea or gravy may be added to the rice or moistened biscuit, which should constitute the chief portion of the diet. Give cod liver oil twice a day in doses of one-fourth to half an ounce in milk, in order to assist the digestion and invigorate the health. At the commencement give $\frac{1}{2}$ grain to four grains of tartar emetic every other morning until six or seven doses have been administered, and if the bowels are confined give 2 to 4 drachms of castor oil until they are opened. A mustard poultice may be applied each successive night to the throat, first clipping off the hair ; but it must be removed whenever the animal appears to be uneasy. Leeches may also be applied to the throat and small blisters to the chest. Mayhew states he has found great improvement result from wearing a very wide bandage, kept constantly wet, and covered with oil silk. Strips of gutta percha or of stout leather will prevent it being doubled up by the motions of the head. A seton can only be inserted with safety by one who is well acquainted with the anatomy of the parts and the distribution of the different veins. Although no one medicine appears to have any specific influence over the disease, the following pills are likely to do good :—

Barbadoes tar, half a drachm to two drachms ; powdered squills, one drachm to four drachms ; extract of belladonna, half a scruple to four scruples ; liquorice powder, a sufficiency. Beat into a mass, and make into twenty pills ; give four daily. Or

James's Powder, one grain to four grains ; Dover's powder, six grains to a scruple ; balsam of Peru, a sufficiency. Make into one pill, and give as before. Or

Extract of hyoscyamus, one to four grains ; powdered ammoniacum and cubebs, of each four to twelve grains ; Venice turpentine, a sufficiency.

If the throat is very sore, hold the mouth open, and blow into it ten grains of powdered alum mixed with four times its weight of fine sugar.

If the cough becomes weaker, less loud, and more short, and also more frequent, ulceration of the windpipe is to be dreaded, in which case it will be humanity to destroy the animal, which may be done instantaneously by dropping five or six drops of prussic acid on the tongue or into the ear.

SNORING.

Causes.—Debility, fat, and sloth.

Treatment.—Restore the strength, if it arises from debility, by giving doses of cod liver oil twice a day ; half an ounce to be given in milk. If fat and sloth have given rise to the habit, stint the food ; give only vegetable diet, and sufficient exercise.

SNORTING.

Cause.—Irritability in a low form of the lining membrane of the nostrils ; worms sometimes the cause.

Symptoms.—The animal stands with his head erect ; draws the air through the nostrils, and produces a series of harsh, loud sounds, which are sometimes continued till the dog falls from sheer exhaustion.

Treatment.—Unless in such cases as arise from the presence of worms, medicinal treatment appears to be uncertain. Let the diet be nourishing, but not stimulating, and the following tonic pills may be given :—Disulphate of quinine, one to four scruples ; sulphate of iron, one to four scruples ; extract of gentian, two drachms to half an ounce ; make into twenty pills, and give three daily ; or shake a little grey powder on the food. Grey powder is a preparation of mercury and chalk, and is used as an alterative for dogs, in doses of 3 to 10 grains.

COUGH.

Causes.—Not rubbing dry after washing ; neglect, and damp kennel.

Symptoms.—Running at the nose, snuffling, watering at the eyes, cough.

Treatment.—In early stages comfort will generally effect a cure. Ascertain the cause, and remove it if possible. If the breathing becomes oppressed, give extract of belladonna combined with James's powder, of each from a quarter of a grain to two grains, according to size. Administer the dose every hour until there is a marked disinclination for food or drink, on which cease giving until the third day, when, if the cure is not complete, the medicine may again be used.   At same time rub a little soap liniment —otherwise known as opodeldoc—along the course of the windpipe and over the chest.   See that the bowels are open, but avoid purging.   Alterative doses of grey powder should be sprinkled on the food.   The dose is from three to ten grains. Turpentine liniment may also be freely rubbed on the sides, throat, and under the jaws.

INFLAMMATION OF THE LUNGS.

Causes.—Often obscure.

Symptoms.—Quickened pulse and breathing, preceded by shivering fits.   The animal is averse to motion, and sits persistently on its hocks.   As the disease proceeds, the breathing becomes worse, and there is an obvious sense of suffocation.   The dog refuses to sit, and obstinately stands.   One of the legs swells, being enlarged by fluid.   At last the animal falls, and finally sinks.

Treatment.—In the commencement of the disease give extract of belladonna combined with James's powder, as recommended in " Cough."   On the second day, if there is no improvement, give every fourth hour from 15 drops to half a drachm of antimonial wine, which must be lessened if vomiting is produced.

In cases which are severe from the first, take from one ounce to eight ounces of blood from the jugular vein, and apply a blister to the sides, first cutting the hair off.   This is done by saturating a rag, folded three or four times, with a solution composed of strong liquor of ammonia one part, and distilled water three parts. Cover the saturated rag with a dry cloth to prevent evaporation. Take it off in ten minutes or a quarter of an hour, lest the effect be too powerful.   Give a dose of castor oil, and the food should be of a vegetable nature.   Homœopathists give from half a drop to two drops of tincture of aconite in water every hour.

DROPSY OF THE CHEST.

Cause.—The result of inflammation of the lungs.

Treatment.—Tapping; the trochar being introduced between the seventh and eighth ribs, close to the breastbone. If the dog shows signs of faintness, stop the operation for a time until strength has been regained. Give a flesh diet, and the following pill three or four times a day :—Iodide of iron, one to four grains; sulphate of iron, two to eight grains; extract of gentian, ten grains to half a drachm; powdered capsicums, two to eight grains; powdered quassia, a sufficiency. This will make two pills; they should be made frequently. Treatment in dropsy is uncertain.

ASTHMA.

Causes.—Inordinate feeding, causing grossness and fatness. Common in old and petted dogs.

Symptoms.—Attack often sudden; appetite not affected, and sometimes becomes much, increased. Piles usually accompany asthma; the coat is in hard condition; the hair falls off in places; nose dry; membrane of the eyes congested; teeth covered with tartar; breath offensive.

Treatment.—Asthma is so seldom curable, and the tendency of the medicines given for that purpose being likely in the end to destroy the general health, it will be for the most part more humane to put an end to the existence of the animal by placing five or six drops of prussic acid on the tongue. At same time, the effects of a very spare diet, from which meat in any shape should be rigidly excluded, should be tried, together with enforced exercise, daily brushing of the skin, and a hard bed. In mild cases, taken in time, these measures will afford a certain degree of relief. The following tonic may also be given :—Extract of gentian, two drachms; sulphate of iron, one scruple. Make into twenty pills, and give one to four in the course of the day, according to the size of the animal.

CHAPTER IV.

DISEASES OF THE INTERNAL ORGANS.

DISEASE OF THE LIVER.

Causes.—Over-feeding and excessive indulgence.

Symptoms.—Mayhew says that when the animal is fat the visible mucous membranes may be pallid; the tongue white; pulse full and quick; spirits slothful; appetite good; bowels irregular; breath offensive; arms enlarged; the rump denuded of hair, the naked skin being covered with a scaly cuticle, thickened and partially insensible. When the animal is thin, the dog may be only emaciated, a living skeleton, with an enlarged belly. It is dull and sleepy; and when disturbed the expression of the countenance is half vacant and wild. The coat does not look positively bad; the membranes of the eye are very pale, but not yellow.

Treatment.—Mayhew recommends a long course of iodide of potassium in solution, combined with the liquor potassæ, as constituting the principal dependence.

Take—Iodide of potassium, two drachms, two scruples; liquor potassæ, one ounce and a half; simple syrup, six ounces; water, twelve ounces and a half.

Mix, and give from half a teaspoonful to a teaspoonful three times a day. This must be persevered in for a couple of months before any effect can be anticipated. The food must be nutritious and digestible, not reduced to the starvation point, but neither must it be given in large quantities. If purgatives are required, give the following pill—Compound extract of colocynth, half a scruple; powdered colchicum, six grains; mercurial pill, five grains; mix. This is a dose for a small dog; a large animal—such as a Newfoundland—would require three times the quantity. A sporting dog, which has been subject to liver complaint, will never be capable of doing work, nor should such an animal be used for breeding purposes.

INDIGESTION.

Cause.—Injudicious treatment with respect to food.

Symptoms.—Dislike for wholesome food, and a craving for hotly spiced or highly sweetened diet; thirst; sickness; propensity to eat string, wool, thread, paper, &c.; peevishness.

Treatment.—Put the animal on a short allowance of easily digested food, with plenty of exercise and a cold bath every morning. A course of the following pills will be useful:—Extract of hyoscyamus, sixteen grains; carbonate of soda, half an ounce; extract of gentian, half an ounce; carbonate of iron, half an ounce. Make into eight, sixteen, or thirty-two pills, according to size, and give two daily. When purgatives are necessary, give castor oil as follows:—Castor oil, four parts; olive oil, two parts; oil of aniseed, a sufficiency; mix, and give with a little powdered sugar.

In old dogs indigestion is usually accompanied with flatulent colic, and also fits and diarrhœa. Give injections of ether and landanum, but avoid bleeding or purging. Be very cautious as to diet, which, while strengthening, should be entirely fluid.

INFLAMMATION OF THE STOMACH.

Cause.—Unwholesome food of some kind.

Symptoms.—Frequent sickness and vomiting; constant thirst, and latterly the draught is ejected as soon as swallowed. The animal frequently licks the fireirons in its desire to touch something cold; it avoids heat; breathing is quick; dislikes motion; stretches itself out either on its chest or its belly; and there i always more or less diarrhœa

Treatment.—Mix half a grain to a grain and a half of calomel with equal quantities of powdered opium, and sprinkle this upon the tongue. Dissolve from one drachm to four drachms of sulphuric ether in water, and give twenty minutes after the powder. Inject ether in water every hour, and from a quarter of a grain to a grain of powdered opium may be sprinkled on the tongue every hour, which must be continued until the sickness ceases, or the animal shows the effects of the opium. Apply a strong embrocation of turpentine to the left side. Give any quantity of cold water; iced, if possible. After the sickness has subdued, give the following pill:—Powdered nux vomica, a quarter of a grain to a grain; sulphate of iron, one grain to four grains; extract of gentian, sufficient to make a pill. This may be repeated every four hours until the stomach is quiet. During this treatment do not give food, and after recovery it should be light, consisting chiefly of cold rice, arrowroot, &c. After the violence of the disease has ceased, give the following pill four times daily for some days:—Extract of hyoscyamus, one grain to four grains; carbonate of soda, three grains to twelve grains; carbonate of ammonia, half a grain to two grains; extract of gentian, five grains to a scruple; powdered quassia, a sufficiency.

ST. VITUS'S DANCE.

This disease, which is usually understood to be a nervous disorder, Mayhew considers is primarily an affection of the stomach and intestines.

Cause.—Follows distemper.

Symptoms.—An uninterrupted catching or twitching of a limb or limbs, which continues day and night, sleeping and waking, standing or lying; otherwise the animal seems well.

Treatment.—Treatment must be chiefly limited to attention to the food, which should consist of a small quantity of rice boiled in some beef tea. Give water, and keep a comfortable bed under the animal, which must be frequently renewed. With careful attention of this kind the animal may get over the disease as he gets older and stronger.

CONSTIPATION.

Treatment.—In ordinary cases a teaspoonful of castor oil is all that is necessary, and a little lard and sulphur mixed answers as a slight laxative as well as a purifier of the blood. If the animal evidently suffers great pain, which will occur in aggravated cases, give tincture of opium, one scruple, and sulphuric ether, four drachms, added to cold gruel; repeat every ten or twenty minutes. Give also copious injections containing linseed oil and some turpentine. If necessary, the hard dung must be removed with the finger, great care being exercised not to rupture the intestine. When relieved, give a dose of castor oil, and attend to the food.

COLIC.

Causes.—Various and uncertain.

Symptoms.—The animal moans; is restless; drags its stomach along the ground, but these symptoms are not attended with purging or vomiting, as in inflammation of the stomach.

Treatment.—In slight cases, rubbing the stomach with a little turpentine or applying a bag of hot salt will give relief. In more serious cases give injections of ether and laudanum in gruel, and doses of the same should be given every half hour. The medicine and injection is made up as follows:—Cold gruel, one quart; sulphuric ether, four drachms; laudanum, one scruple. This is sufficient for a large dog; for a small one an ounce of the mixture will be sufficient for a dose or injection. Give a dose of castor oil, and be careful in the feeding afterwards.

WORMS.

Symptoms.—Uncertain appetite; variable spirits; unthrifty coat; leanness; colic; fits; whining; trying to bite under the tail, and dragging the rump along the floor.

Treatment.—Give, when fastiug, oil or spirit of turpentine in doses of from half a drachm to two drachms in a little olive oil. The areca or betel nut is also a useful remedy, powdered, and given, when fasting, in doses of 30 grains to 2 drachms, in milk or gruel.

INFLAMMATION OF THE BOWELS.

Cause.—Unwholesome food; exposure; over-exertion; neglected colic.

Symptoms.—Colic aud coustipatiou, succeeded latterly by diarrhœa; extremities cold; pupil of the eye much dilated; breath hot, and nose dry. Tail is firmly pressed dowuwards; urine scauty, and light-coloured; tongue rough; great thirst; but no appetite. When diarrhœa sets in, recovery is hopeless.

Treatment.—Doubtful as to result. Give a hot bath of ninety degrees, and afterwards cover up with several hot blankets. Give a turpentine injection, and calomel and opium, as in inflammation of the stomach; also doses and injections of ether and laudanum, as iu colic, followed by castor oil. Apply a mustard poultice to the belly, followed by fomentations with warm water, and occasional fomentations of hot turpentine. The food, after the disease has abated, must consist only of beef tea and rice.

DIARRHŒA AND DYSENTERY.

Causes.—Bad food; excessive fatness and want of exercise; colic.

Symptoms.—Sickness, and the animal does not, as in other cases of sickness, consume what has been thrown off the stomach; thirst; breath offeusive; appetite raveuous; piles; dung liquid, and sometimes black, mixed with mucus and blood.

Treatment.—Give the following medicine every two hours until the desired effect is produced; the amount of the dose must depend on the animal's size:—Powdered opium, half a grain to two grains; powdered chalk, five grains to a scruple; catechu, two grains to half a scruple; liquor potassæ, half a drachm to two drachms; powdered ginger, three to twelve grains; powdered carraways, three to twelve grains; powdered capsicums, one to four grains. Feed the dog on beef tea thickened with rice. Warm fomentations should be applied repeatedly to the belly, covering up afterwards with hot blankets. Mayhew recommends the following ointment to be smeared over the anus and introduced up the rectum by means of a penholder:—Powdered camphor, mercurial ointment, elder ointment, equal parts of each. Clean the anus and root of the tail several times daily with a wash consisting of an ounce of chloride of zinc to a pint of distilled water.

The following recipes for pills to be administered in diarrhœa and dysentery are given by Gamgee :—

| Mercury with chalk, four grains; Dover's powder, five grains; confection of roses, sufficient. Mix.

| Sulphate of copper, four grains; extract of opium, two grains; extract of gentian, sixteen grains. Mix, and make into six pills. Useful in chronic diarrhœa in dogs.

STOPPAGE.

Causes.—Swallowing sticks or stones thrown in sport for the animal to catch; also swallowing sharp pieces of bone and other hard materials.

Treatment.—Give as much food as the animal will eat, and half an hour after administer a large dose of antimonial wine, say two to four ounces, and take the dog out for exercise. This will cause vomiting, by means of which the foreign substances may be ejected. If this should not be the case, administer a quantity of thick gruel, and afterwards some more antimonial wine—a smaller dose; but should the stone, or other article, reach the bowels, treatment is generally useless.

N

PARALYSIS OF THE HIND EXTREMITIES.

Causes.—Mayhew considers this affection is caused by a disordered state of the stomach, arising from over-gorging and excessive fatness.

Treatment.—Give a few of the following purgative pills:— Compound extract of colocynth, half a scruple; powdered colchicum, six grains; mercurial pill, five grains. This is for a small dog; and the quantity must be increased from two to three times if the dog is large. Follow with castor oil, prepared as follows:—Castor oil, four parts; olive oil, two parts; oil of aniseed, sufficient. Add a little pounded sugar, and mix. Attend carefully to diet, which should be spare.

VOMITING.

Causes.—Disordered stomach.

Treatment.—When a dog is troubled with continued vomiting, give the following pill every four hours until the desired effect is produced :—Powdered, vomic nut, one-fourth of a grain to a grain; sulphate of iron, one grain to four grains; extract of gentian, sufficient to make into one pill.

CHAPTER V.

DISEASES OF THE SKIN.

MANGE.

Causes.—Arises from attacks of an insect, brought on by want of ventilation in and uncleanliness of kennel ; also injudicious feeding, such as a diet chiefly or wholly composed of flesh, &c.

Symptoms.—The hair comes off in patches ; the skin is dry and scaly, and is wrinkled.

Treatment.—Mayhew strongly recommends the following ointment, which must be well rubbed in on the skin, and not merely smeared over the hair :—Sulphur, one and a half ounce ; oil of juniper, half an ounce ; resin ointment, three ounces. This is to be well rubbed in one day, and washed off the next, and repeated as often as necessary. The food must be strictly attended to, and flesh meat entirely forbidden. Boiled rice is the best food ; and if the animal refuses to eat it, let him alone, and offer it fresh every day until he does eat it. Fasting will do him no harm. Give a dose of castor oil, prepared as directed in " Paralysis of the Hind Extremities," and repeat the dose for three days. Give a cold bath every morning. The following stimulating liniment may then be applied every day where the hair is wanting :—Oil of turpentine, oil of tar, and linseed oil, of each four ounces. Mix. After being applied for a week, add another ounce of turpentine to the liniment, and use until the skin is reduced to its natural state.

Gamgee gives the following lotions as useful for red mange in dogs :—Solution of potash, half an ounce ; dilute hydrocyanic acid, two drachms ; water, seven and a half ounces. Mix.

Sulphate of zinc, one drachm; carbonate of soda, two drachms; water, eight ounces: mix.

Liquor arsenicalis given on a piece of meat, or diluted with water, is much recommended in cases of mange. The dose consists of one to two drops: three to be given in the course of the day; and to be increased a drop each day until the dog loathes his food and has a running from his eyes, when the medicine is to be discontinued for three or four days, and then resumed, but not increased beyond the point which threw the animal off his food. The doses are given morning, noon, and night.

The cure of mange is for the most part a slow process, and the steps taken must be steadily persevered in.

VERMIN.

Fleas.—Remove the bed, and sluice the kennel with boiling water, and afterwards paint it over with spirits of turpentine. Wash the dog with the yolk of eggs beaten up, and having a teaspoonful of spirits of turpentine added to each egg, also sufficient water. Let the bed consist of clean, yellow deal shavings. When the fleas are not very numerous, a little powdered camphor rubbed into the hair will often prove effectual in banishing them.

Lice.—These frequently cover the body, and are also found in large numbers about the eyes and lips. Apply castor oil to the body until every part of the hair is completed saturated. Shut him up for a day, and remove the oil with a wash made of the yolk of eggs and water. The castor oil will likely act on the bowels, which will be beneficial.

CHAPTER VI.

SPECIAL DISEASES.

~~~~~~~~~~~~~~~~~~~~~

RHEUMATISM.

Causes.—Over indulgence in food.

Symptoms.—Bad temper; the animal howls when touched, and even without being touched; but on pinching the skin the animal seems to like it. The teeth are covered with tartar, and the breath is most offensive.

Treatment.—Limit the food, and give only a little rice and gravy, once a day; if the animal refuses to use it, take the food away, and offer some more of the same kind next day, freshly made up; if the dog still refuses to eat, take it again away, and repeat until hunger compels him to eat. Give the purgative pills and castor oil as recommended in " Paralysis of the Hind Extremities," and rub the back, neck, and belly, three times each day, with the following liniment—turpentine, laudanum, and soap liniment, of each one ounce; tincture of capsicum, one drachm; mix. Gradually increase the quantity of turpentine in the liniment to one and a half ounce, when the liniment may be applied only twice daily. When all signs of pain are gone, reduce the quantity of turpentine to one-third, and apply only once a day. At the expiration of a month leave off all treatment for a week, but repeat if necessary.

PILES.

Causes.—Various, but generally disordered digestion.

Symptoms.—A watery discharge from the fundament, which

protrudes, and the edges become covered with cracks and ulcers.

Treatment.—Regulate the diet strictly. Give the animal lean meat, from two ounces to two pounds each day according to size, divided into four meals; let the animal have also plenty of exercise and a cold bath daily. Smear the outside of the fundament with some of the following ointment, and also insert some of it by means of a wooden pen-holder, three times daily:—Camphor, two drachms; mercurial ointment, one drachm; elder ointment, one ounce; powdered opium, one ounce; mix. The dog will resist, but the ointment must be applied, and after three or four days the pain, which caused the animal's opposition to treatment, will cease. If there is a strong smell from the parts, moisten the fundament at each dressing with chloride of zinc, largely diluted with water. If purgatives are required, give a little olive oil, but nothing else.

FOOT LAMENESS.

Cause.—Common in game dogs which are brought out for autumn work without previous preparation.

Treatment.—Mayhew states he adopts the following treatment:—Get a basin of tepid water and a soft sponge, and well wash the injured foot. When every particle of dirt is removed, apply to the dried sore surface a lotion composed of two grains of chloride of zinc to one ounce of water, with one or two drops of the essence of lemons. Having thoroughly washed the foot with this lotion, soak some rags in it, and wrap them round the injured foot, fixing over all a leather or gutta percha boot. The animal must be allowed to rest for a few days, and then brought to work with caution.

A writer in the *Field* recommends the following treatment for foot lameness:—Take one and a half pounds of sulphate of iron (green vitriol), one pound of alum, two ounces of verdigris, and one ounce of sal-ammoniac. These are to be well powdered, and put into a glazed earthen pipkin, that will hold a quart or three pints; put it on a moderate fire, which may be increased until the contents boil up two or three times; then take them from the fire, and set them to cool for six or seven hours; break the pipkin, and take out the stone; it must be stirred with a piece of wooden lath all the time it is on the fire. Pound a piece of the stone, about the size of a walnut, and melt in a quart of rain or soft pound water; shake the bottle well when used, pour some of it into a cup, soak a piece of linen well in it, and apply the linen in eight or ten folds to the foot, keeping it constantly wet, for which purpose it ought to be covered over with a piece of oiled silk.

DISTEMPER.

Causes.—Cannot be well ascertained. Some suppose it arises from contagion, but Mayhew doubts this on apparently good grounds. Others assert it is a disease which all dogs must take, as children do certain diseases; but we and many others have reared dogs which never had distemper.

Symptoms.—Dulness and loss of appetite, purging or vomiting, unusual moisture of the eyes, and a short cough, are usually the earliest signs. As the disease progresses the animal is constantly shivering, and seeks warmth; it trembles violently, especially when taken hold of; the bowels are constipated; a thick discharge flows from the eyes; the white of the eye and the inside of the upper lid become covered with numerous small red vessels, bright in colour, and having their course towards the centre of the eye. A glairy, yellow fluid covers the nostrils, and the breathing is accompanied by an unusual sound, which may be detected by applying the ear to the head. The cough becomes severe, frequent, and spasmodic, and after each fit a quantity of yellow frothy liquid is thrown off the stomach. The matter vomited has an offensive smell; the nose is harsh and dry; the coat staring; skin hot, and paws warm; the pulse quicker than it was at first. The symptoms become aggravated as the disease advances, and the eyelids are glued together with thick matter; the nostrils become filled with matter, which impedes the breathing, and the animal endeavours to remove it by using its paws. The cough continues, but it is more irregular than it was in the previous stage; the shivering is constant; ulcers appear on the hips; the animal utters short, sharp cries; lies on its side; diarrhœa sets in, which is followed by death. Fits also frequently occur at intervals, previous to which the appetite becomes great. Yellow distemper is jaundice added to the original disease. The disease lasts from three to six weeks, but so long as the animal can move about there is a chance of recovery. The worst symptoms are when the tongue becomes coated, discoloured, and red and dry at its tip and edges, and when the animal falls down, and lies on its side.

Treatment.—Mayhew places great reliance on strict attention to the diet, which should consist of boiled rice, with a little broth from which the fat has been skimmed off, or bread and milk given cold. Meat of all kinds, sugar, butter, tea, or hot liquor must be rigidly withheld. Water, constantly changed, must be the only drink, and the animal may be covered with a piece of blanket thrown loosely over it. Let the bed be fresh made up each day, and plenty of sweet straw is the best material. It must, of course, be in a dry place, not exposed to draughts. In the early stage give an emetic three successive mornings, and the best is half a teaspoonful to a dessert spoonful of antimonial wine. On the fourth day give a gentle purge, consisting of castor oil, four parts; olive oil, two parts; oil of aniseed sufficient. Mayhew gives at the same time the following pills:—Extract of belladonna, six to twenty-four grains; nitre, one to four scruples; extract of gentian, one to four drachms; powdered quassia, a sufficiency. Make into twenty-four pills, and give three daily. Even should the symptoms disappear under this treatment, strict attention to diet must be persevered in, a course of tonics adopted, such as the following:—Disulphate of quinine, one to four scruples; sulphate of iron, one to four scruples; extract of gentian, two to eight drachms; powdered quassia, a sufficiency. Make into twenty pills, and give three daily. Mix ten or twenty drops of the liquor arsenicalis with one ounce of distilled water, and add a little simple syrup. Of this give a teaspoonful three times daily in a little milk, and continue for three or four weeks, after which, if no further symptoms manifest themselves, the animal may be considered cured.

If, however, the disease has got into an advanced stage before treatment is resorted to, Mayhew states that in such cases he neither purges nor uses emetics, but if the fundament is full of hardened dung, he gives an injection, composed of cold gruel, one quart; sulphuric ether, four drachms; and laudanum, one scruple. This is sufficient for a large dog; for a small one, an ounce of the fluid will be sufficient. Neither does Mayhew interfere with the cough, or with the state of the eyes. In fact, he is totally opposed to bathing the eyes, or the application of lotions, powders, bleeding, blistering, and setoning, as means of relief, but attends strictly to diet, giving sufficient to prevent the animal from starving, without cramming it; little and often being the rule. As medicine he recommends the following pill to be given thrice daily:—Extract of belladonna, one to four grains; nitre, three to eight grains; James's powder, one to four grains; conserve of roses, a sufficiency. This is sufficient for one pill. When a quantity of pills are made up at once, one drop of the tincture of aconite may be added to each four pills. If diarrhœa sets in, give ether and laudanum both by the mouth and by injection. One pint of cold gruel, two ounces of sulphuric ether, and four scruples of tincture of opium: of this from half an ounce to a quarter of a pint may be administered as an injection; and from one to four tablespoonfuls as a dose by the mouth, in a little syrup; to be given every second hour, and if colic appears, every hour or oftener. Should the diarrhœa continue and the colicky pains disappear, add five to twenty drops of the liquor potassæ to every dose of ether given by the mouth, which should then be every third hour, and give the following astringent pills:—Prepared chalk, five grains to one scruple; powdered catechu, two to eight grains; powdered ginger, three to ten grains; powdered carraways, three to ten grains; powdered capsicums, one to four grains; confection of roses, a sufficiency. When the dung has an offensive smell, a wash consisting of two ounces of the solution of chloride of zinc to a pint of cold water is made use of to cleanse the parts about the roots of the tail.

When the symptoms indicate approaching fits, the following medicine must be administered thrice daily, but if it produces sickness, the dose must be reduced to half the quantity:—Grey powder (mercury with chalk), five grains to one scruple; ipecacuanha, one to four grains. Also, tincture of hyoscyamus, one part; sulphuric ether, three parts; mix this with cold syrup, in the proportion of ten ounces of syrup to one ounce of the medicine. Give an ounce of the mixture every hour to a small dog, and four ounces to a large one. Give an injection of the solution of soap, and afterwards, every hour, an injection, an ounce to four ounces, according to size, of the sulphuric ether, hyoscyamus, and soap mixture. Shave the hair from the back part of the head, and apply one to four leeches to the part. Apply a small blister to the nape of the neck, composed of equal parts of liquor ammonia and camphorated spirits. Remove the hair, and after saturating a piece of sponge with the blistering compound, apply it from five to fifteen minutes. If necessary to repeat it, let it be applied a little lower down towards the shoulders, but never on the same place. If the blister rises, some hope of recovery may be entertained; but Mayhew entertains little hope of an animal affected with distemper when the disease reaches the stage in which fits occur.

BITCHES EATING THEIR YOUNG.

Cause.—Mayhew attributes this unnatural propensity to an affection of the brain, induced probably by persecution, chastisement, or some other species of annoyance.

Treatment.—When a bitch has devoured her young, give an emetic, followed by an aperient, and after it has operated give tonic medicines. Let the food be mild, and to dispose of the milk, rub in camphor dissolved in sweet almond oil. Next time the bitch is breeding, let her be kept very quiet, give a gentle aperient dose, and let her be plentifully supplied with water, frequently renewed, to drink.

WARTS.

Treatment.—Dress the warts every day, for three or four days, with an ointment made of one part of arsenious acid to four parts of lard, which will effect the removal of the warts.

ADMINISTRATION OF MEDICINE TO DOGS.

Physic may sometimes be administered to a dog in a thin slice of meat, or when mixed with butter, or milk, or any food of which the animal is usually fond. When it is necessary to administer it otherwise, take the dog, if small, into the lap, the person who is to give the medicine being seated. Grasp the skull with the left hand, and press the thumb and fore finger against the cheeks so as to force them between the posterior molar teeth. Drop the pill as far as possible into the mouth, and push it down the throat with the fore finger of the right hand. When the finger is withdrawn, the jaws ought to be clapped together, and the attention of the animal diverted. When the animal proceeds to lick the nose and lips, it is a sign the substance has been swallowed. Some medicines may also be given by pouring it over the throat with a spoon, allowing the animal time to swallow it properly.

In the case of large dogs it may be necessary to tie the mouth before administering medicine. This is done by introducing a rounded piece of wood into the mouth, which is secured by tapes ; or, two pieces of tape are taken, one of which is passed behind the tusks of the upper, and the other, in like manner, upon the lower jaw. An assistant pulls on the tapes, which forces the mouth open, and it is then held in that position. Mayhew prefers the stomach pump.

PART VI.

DISEASES OF POULTRY.

GAPES OR PIP.

Causes.—Small worms in the throat, caused by impure water, filthy and damp houses or coops.

Treatment.—By constantly keeping a small lump of camphor, about the size of a field bean, in the water which the fowls drink, gapes will be prevented. When fowl are affected, give a dose of camphor, size of a garden pea. A correspondent of the *Journal of Horticulture* (Poultry Chronicle) recommends the following:— Bole armenian, 20 grains; spirits of tar, 12 drops; cochineal, 1 oz.; mix, and divide into pills the size of a peppercorn, and give on the first symptoms appearing. Attend to warmth, fresh air, and nourishing food. Wheat is excellent food for fowls any way out of order.

ROUP.

When taken at the outset, roup may be cured by feeding twice a day with stale crusts of bread soaked in strong ale: dry housing and cleanliness are also indispensable. Remove the infected from the healthy birds, and feed both freely as above. Wash the heads once or twice a day with weak vinegar and water. When young chickens have both eyes closed, kill them at once.

DIARRHŒA, OR SCOURING.

Cause.--The too abundant use of relaxing food.

Treatment.—Cayenne pepper or chalk, or both, mixed with meal or boiled rice, will check the disease.

BALDNESS.

Rub the part with sulphur ointment.

LEG WEAKNESS.

This is generally caused by the size and weight of the body being more than the legs can bear. It is shown by the bird resting on the first joint.

Treatment.—Tincture of iron, five drops to a saucer of water, is beneficial.

WASTING.

Fowls sometimes waste away without any apparent disorder. In such cases a teaspoonful of cod liver oil will often be found a most efficacious remedy.

APOPLEXY.

This disease is caused by high feeding, and is most common among laying hens, which are sometimes found dead in the nest; the expulsive efforts required in laying being the immediate cause of the attack.

Treatment.—The only hope for cure consists in an instant and copious bleeding. The largest of the veins on the under side of the wing should be selected and opened in a longitudinal direction. Press the thumb on the vein between the incision and the body, to ensure a free flow of blood. Light food and quiet is essential after the operation.

DROPPING OF THE CROP.

Causes.—A rupture or relaxation of the skin; at other times a fevered state of the body, which causes intense thirst.

Symptoms.—A swelling in the breast; great thirst.

Treatment.—Confine the bird, and give only a small allowance of water, which should have vinegar or wormwood mixed with it. A hen so affected will go on laying, but the lump will remain in front.

CATARRH.

Symptoms.—Gurgling noise in throat, and occasional cough and expectoration.

Treatment.—Pour six drops of Powell's balsam of aniseed into a teaspoonful of port wine, and give this as a dose at night. Next night but one repeat the dose, if necessary.

CRAMP.

Causes.—Bad or improper food.

Symptoms.—Inability to stand up.

Treatment.—Purge freely with castor oil, a tablespoonful to the dose, and repeat it every twelve hours until the bird is relieved. If it seem to suffer from the purging, give him bread steeped in ale or in wine.

FOWLS PICKING EACH OTHER.

Fowls, especially in confinement, are always liable to pick each other's feathers. All crested breeds, such as Polish, Houdans, and

Crevecœurs, are very prone to it. It generally comes from vitiated appetite, and can be cured by a good teaspoonful of oil and low diet. Worm diet, from its power to cure this habit, generally keeps birds at liberty in health; the want of such in confinement and heat of blood cause it; and as most people consider giving meat food will cure the affliction, generally " fuel is added to the fire" by over-heating the already feverish state of blood.

COCK'S COMB BLEEDING.

Cauterise the wound freely with lunar caustic.

SWOLLEN FEET AND LEGS.

Dorkings are very liable to this affection. If an old bird is attacked, kill it at once, as a cure is hopeless; if a young bird, feed generously on cooked meat, oatmeal mixed with milk, and hard boiled egg. Keep the bird confined, with hay to tread on. Let it have a large sod of growing grass fresh cut every day. If the swelling extends above the knee joint, and the flesh of the thigh, close to the knee, is uneven on the surface, and feels as if filled with air cells, kill the bird at once.

LOSS OF FEATHERS.

Feathers will, at all times, drop off fowls when not moulting, rendering them often nearly naked. This arises from a disease akin to mange. Give alteratives, such as sulphur and nitre, mixed with fresh butter; change the diet, and attend to cleanliness. An ointment composed of one part of sulphur to sixteen parts of lard may also be rubbed in over the body. The difference between this affection and moulting is, that in moulting the feathers are replaced by new ones as soon as cast, while in this disease the fow becomes bald.

TOBACCO A CURE FOR SICK POULTRY.

A correspondent of the *Gardeners' Chronicle*, February 15th, 1869, gave the following account of his experience regarding the effects of tobacco as a remedy for ailments in poultry. We may mention that his statements as to the curative effects of tobacco were subsequently corroborated by other correspondents of that journal:—" Speaking with the wife of a working bailiff who had been a successful raiser of fowls, I asked what plan she adopted when they were sickly, she quickly made answer—' I give them a " quid" of tobacco.' This reply so acted on my risible faculties that I could not follow up the conversation; but she further stated—' I have adopted the plan with success for ten years.' I then inquired why she gave it, and the quantity administered, to which she replied—' I had noticed than when my husband was mopish and out of sorts that if he took a large ' quid' of tobacco he soon came round; and the thought occurred to me that it might relieve my fowls, which it always does; so whenever I see any of them out of sorts, I give them a piece of tobacco as large as from the end of my thumb to the first joint.' You can judge my surprise as a medical man, when I state that I have seen a like quantity destroy life in a human being. Now for the sequel. In the autumn of last year I purchased some prize fowls, and one of them a month since became sickly. I gave the old woman's remedy—a piece of tobacco the size of the first joint of my thumb (*i.e.*, 30 grains). It had a most speedy and singular effect upon it; in two minutes there was a little staggering, accompanied by a peculiar twitching of the tail, which gradually became straight with the back, and ultimately trailed on the ground; in twenty minutes the fowl appeared quite well, and has continued so. This morning my servant, as usual, let the fowl out and gave them some barley, but the cock bird appeared very sickly and disinclined to eat. He stood with his mouth slightly opened, and wings hanging down; in fact, what the old woman termed 'out of sorts.' I offered him some pieces of bread—my constant practice;

O

hut he took little notice, or of the hens whilst eating it. As this state had lasted three or four hours, I looked down his throat, which appeared healthy, and he had nothing in his crop. I then gave him the 'quid' of tobacco, *i.e.*, 30 grains. In two or three minutes he appeared very weak, and his tail began to drop slightly; he then sat down under a tree, aud remained quiet about five minutes. I then walked to him, when he got up and in a few minutes commenced pecking some corn, and in a quarter of an hour from the first taking of the tobacco he appeared quite well and began to crow most lustily, although he had not made the slightest effort before during the morning, which was very unusual, as he frequently crows when well. To see him now, twenty-four hours after the dose of tobacco, performing his accustomed duties, no one would scarcely believe he had taken so potent a remedy. I do uot profess to give the *modus operandi*, but as it acts like a charm it is worth knowing."

PREVENTION OF DISEASE IN POULTRY.

Give finely chopped onions, mixed with a little meal, two or three times a week; give as much as they will eat. Give toast dipped in ale, and as a tonic put some rusty nails in the water pan.

—————

NOTE.—As a general rule, when anything is amiss with poultry, it is advisable to make them ready at once for the pot. Treatment may be desirable in the case of more than usually valuable fowls, but it is frequently unsuccessful, aud renders the fowl unfit for use.

PART VII.

CHAPTER I.

WOUNDS, FRACTURES, ETC.

When a serious injury happens to an animal, a competent veterinary surgeon should be called in immediately, as it is only such who can treat matters of this kind in a proper manner. There are some kinds of injury which, when they occur, are beyond treatment, and the most humane plan is to put the poor animal out of pain at once. At the same time, the following remarks are offered, as they may be useful where a competent professional veterinary surgeon is not within call, and also with the view of enabling persons to meet emergencies, until the arrival of a veterinarian.

WOUNDS.

These are of various kinds, and call for different treatment.

Incised Wounds.—These are caused by a sharp instrument, which has divided the skin. These commonly heal by adhesion. Bathe the wound with cold water, to stop the bleeding, and bring the parts together by divided sutures or stitches. When these begin to drag, cut them across. Dress with lint steeped in friar's balsam, or collodion and glycerine; of each one ounce may be applied, the hair being first removed from about the part. Mayhew recommends that after copious suppuration has been established, the part should be frequently bathed with the solution of chloride of zinc, one grain to the ounce of water.

Lacerated Wounds.—When wounds are much torn, they are called lacerated. Cleanse the wound, by allowing lukewarm water to fall in a stream over it from the mouth of a vessel; or water may be allowed to trickle over it from a sponge pressed above the wound, but not on the wounded surface itself. Lacerated wounds seldom bleed much; but if there is much bleeding, apply a bandage, which should be adjusted evenly, but not so tight as to obstruct circulation. Apply a poultice made of one-fourth brewer's yeast and three-fourths of any coarse meal. As soon as the inflammation has been subdued, apply lint steeped in cold water, and covered with gutta percha. In all cases of wounds keep the diet low for a few days; but after inflammation has ceased, the diet should be liberal, in order to sustain nature. Spooner recommends, in cases of lacerated and contused wounds, when a healthy surface has been obtained by poulticing, &c., to stimulate the wound daily with a little compound tincture of myrrh, and to protect it from the atmosphere by an astringent powder, such as the following :—Powdered prepared chalk, one ounce; powdered alum, one drachm; powdered Armenian bole, one drachm; powdered sulphate of zinc, one scruple; to be well mixed together. Mayhew's directions consist in first giving a drink composed of sulphuric ether and laudanum, of each one ounce; water, half a pint. Repeat the medicine every quarter of an hour, or till shivering has ceased, and the pulse is healthy. A poultice of brewer's yeast and coarse meal, as above, is to be applied; or a lotion consisting of tincture of cantharides, one ounce; chloride of zinc, two drachms; water, three pints.

When the slough has fallen, apply frequently a solution of chloride of zinc, one grain to the ounce of water, and regulate the food by the pulse. A slight scab over a wound is useful; but if it becomes thick and unyielding, it will do mischief, by preventing the confined matter from escaping.

Contused Wounds.—In slight cases dress the part with the ointment of the iodide of lead, one drachm of the iodide to an ounce of lard. In large wounds apply cold water, and divide the skin every eighth inch across the whole swelling, and then bathe the part with the chloride of zinc lotion, as above. Let the diet be liberal, but not heating.

Punctured Wounds are dangerous, as they are apt to form abscesses. Keep the opening free and also clean, by frequent washing and dressing, and in some cases it will be necessary to insert a pledget of tow, smeared with ointment of resin or ointment of carbonate of zinc.

Indolent Wounds are those which heal slowly and imperfectly. In such cases caustic may be applied to the wound, or the neighbouring parts may be lightly blistered. Strengthen the health of the animal by tonics and liberal diet.

When there is an excessive quantity of proud flesh, apply sulphate of copper or zinc, nitrate of silver, or alum.

In simple wounds, where there is no inflammation, the best dressing is the ointment of carbolic acid, spread on tinfoil, and applied to the wound. The bandages which keep the dressing in its place should not be removed for some days; but the dressing must be removed from time to time, until the wound has been cleansed, and begun to heal. Remove the hair from the part before applying the dressing.

In all cases of wounds perfect rest is indispensable; and in some it may be necessary to put on a cradle, in order to prevent the animal gnawing the wound with its teeth.

/ FRACTURES. /

In most cases, when fracture of any of the bones occur in the horse it is useless to attempt treatment, not because the bones of a horse, when broken, will not re-unite; but from the fact that as the value of a horse depends on his being able to move about freely, a horse lamed in consequence of fracture is of very little service to his owner. Where it is only some of the minor bones that are broken, it may, however, be advisable in the case of valuable brood mares or stallions to attempt as perfect a cure as circumstances will admit of.

There is one common kind of fracture which does not interfere much with the usefulness of the animal. This is fracture of the hip bone, which when it happens causes the horse to become what is called "down in the hip." This particular kind of fracture is caused by a fall, or by the horse making a sudden rush in or out of the stable door, and the symptoms are drooping of the quarter, a small crackling noise, and one side of the haunch bone being less prominent than the other. The treatment, when a case of this kind occurs, consists in keeping the animal perfectly quiet, and prevented from lying down. Nature will do the rest, and in three or four months the animal will be fit for work. The food must be of a cooling nature, but nourishing at the same time.

Fractures rarely occur in cattle; in the sheep and dog they are more frequent, but they heal with comparative facility in these animals, unless the bone is broken in several places. A simple fracture is the breaking of a bone across in two pieces.

As it is only a professional man who can detect the exact extent of the injury in cases of fracture, and determine whether it is worth while trying treatment or not, it is needless going into details, beyond stating in a general way what should be done in cases of simple fracture.

Colonel Fitzwygram, in his "Horses and Stables," gives the following directions on this point:—

"The first object is to ' set' or briug together the broken ends of the bone as soon as possible. When the boues can be properly replaced at once, the fractured surfaces are thereby prevented from grating against each other, and from irritating the adjoining parts; in such cases there may be no serious amount of swelling to interfere with the due and continued apposition of the parts and the commencement of the process of healing or union.

"When the bones are thus adjusted, the next thing, if possible, is to keep them in their places. This is ofteu a very difficult matter, and needs expertness and ingenuity. Splints padded with tow and bandages and strips of adhesive plaister may be used, and in some cases the horse may be slung with the view of taking the weight off the part affected.

"The starch bandage is a very useful application. It is formed by soaking iu thick starch mucilage strips of linen, which may be placed, one over the other, in layers, until a firm splint is formed. Additional support may be given by a well adjusted wooden splint outside all. Pads of fine tow will be found useful in preventing undue pressure on particular parts.

"Before applying the starch bandage the part should be oiled, and then a piece of tape should be placed longitudinally over the fracture. The end of the tape should be left hanging out, so that if excessive swelling of the limb takes place, the bandage may be ripped open, and taken off, and dipped in warm water, and then reapplied without losing its shape or mould. Great ease and relief will, in such cases, be given to the patient by this change. Longitudinal slits in the bandage will also be found to give ease.

"Gutta percha softened iu warm water, and moulded to the shape of the bone, likewise forms a serviceable bandage.

"It must be borne iu mind that it is not merely a support, but an easy support, which is required. Pressure cannot be borne. Inflammation, but not repair, will follow on uneasiness; and then the patient will be rendered irritable, and by his movements will certainly frustrate all our efforts at cure.

" When the parts are much swollen and tender, any undue in-flammatory action must, as a preliminary step, be reduced by hot fomentations, as far as possible ; or where the mischief is circum-scribed or almost superficial, by wet cloths kept constantly moist with cold water or refrigerant lotion."

Mayhew in his work on " Dogs," when treating of fractures, says, " It is well to bathe the fractured limb, splints and all, in the following lotions :—

Lotion for the Leg before the swelling has commenced.

Tincture of arnica	... ...	1 drachm.
Water ...	... ...	1 ounce.
Essence of lemon	... ...	A sufficiency.

To be applied frequently.

Lotion to be used when swelling is present.

Tincture of aconite	... ...	½ scruple.
Water ...	... ...	1 ounce.
Essence of aniseed	... ...	A sufficiency.

Lotion to be applied after the swelling has subsided.

Chloride of zinc	... ...	1 grain.
Water ...	... ...	1 ounce.
Essence of aniseed	... ...	A sufficiency.

CHAPTER II.

OPERATIONS, APPLICATIONS, ETC.

BLEEDING.

Bleeding is now much less frequently resorted to than it was some years ago, as it has become a recognized principle that, in the generality of cases, it is much better to keep up the strength of the animal than to reduce it. Colonel Fitzwygram says that "when it is thought necessary to employ it, blood enough should be taken to produce a marked alteration in the character of the pulse. The blood should be drawn in a full stream, so as to produce the effect as quickly as possible. If bleeding is resorted to at all, it should be in the very early stage of the disease, before the strength fails. Bleeding, however, is not a safe remedy. As a general rule it should be avoided; and if there is any doubt as to the advisability, it is always safer not to bleed."

The practice of bleeding young cattle in spring is not to be recommended, if the animals are otherwise in good health.

The operation is best performed in the jugular or neck vein. Either the fleam or the lancet may be used. When blood is to be drawn, the animal is blindfolded on the side to be operated upon, and the head held to the other side; the hair is smoothed along the course of the vein by the moistened finger, the point selected being about two inches below the angle of the jaw. When the operation has been performed, and a sufficiency of blood taken, the edges of the wound must be brought accurately together, and kept so by a small sharp pin being passed through them, and

retained by a little tow. It is essential in closing the wound to
see it is quite close, and that no hairs or other foreign bodies
interpose. For a time the head should be tied up, and care taken
that the horse does not injure the part.

CASTING.

This is the term used for throwing down a horse or bullock, and
so keeping it for the purpose of performing an operation. In the
case of the horse this is done by means of hobbles, side-lines,
barnacles, and trevis. Hobbles are strong straps and cords,
particularly arranged, which are first attached to the feet and
then suddenly drawn together, so that the animal must fall—the
fall being regulated by one man at the head and another at the
haunch. Even when most skilfully performed, many accidents
have occurred to man and horse, from the act of falling and the
struggles of the animals after it. In the case of the ox, take a
long rope, double it, and tie a knot about a yard from the end,
so as to leave a noose of sufficient size to go round the bullock's
neck; which being put on, the two ends are to be brought
between the fore legs, and round the hind pasterns, then back
again and through the noose. By standing in front of the animal
and drawing up the ropes quickly, so that the hind legs are drawn
towards the chest, it is easily thrown down. While in this
situation the ropes must be properly secured, and any operation
may then be safely performed.

CASTRATION.

Colts.—Late in spring or early in the autumn, when the
weather is cool and dry, is the best time to perform this
operation. The colt should be at least a year old before he is
castrated. Keep the animal shut up for twelve hours without
food previous to the operation, and shelter him afterwards if the
heat is great, or the weather should be rainy; otherwise he will

be better running in the field, as moderate exercise is of advantage to him.

There are different modes by which the operation is performed. The animal being cast, the operator, according to one mode, opens the scrotum or bag, and drawing out the testicles, sears them off with a hot iron. Youatt recommends this mode as being simple and successful.

Another mode of castrating colts is by means of the caustic clam. This consists in opening the scrotum and compressing the cord between two pieces of wood, constructed for the purpose, so as to compress the parts as if in a vice. A caustic preparation is smeared on the wound, which prevents excessive bleeding. The clams are left on until the testicle drops off, or is removed by the operator. The caustic paste used for the clams is composed of one part of bichloride of mercury (corrosive sublimate) to eight parts of lard; but the clams invented by Professor Varnell, of the Royal Veterinary College, supersedes the necessity for using this or other dressings of a similar nature.

Calves, &c.—The best time for performing the operation on bull calves is when the animal is about three weeks old. Deprive the calf of a meal before operating, and keep him quiet and cool afterwards. The manner of proceeding is as follows :— Grasp the scrotum in the hand between the testicles and the belly, and make an incision on one side near the bottom, sufficiently deep to penetrate through the inner covering of the testicle, and long enough to admit its escape. The testicle immediately bursts from its bag, and is seen hanging by its cord. Tie a piece of small string round the cord so as to compress the blood vessels, and then divide the rest of the cord a little below the string. The other testicle is proceeded with in the same way. The cord will be drawn back into the scrotum, and the ends of the string will hang out through the wounds. In about a week they will drop off, and the wounds will soon heal. Old bulls may be castrated by means of clams, as stated in the preceding section.

Lambs.—The operation may be performed when the lamb is from a week to three weeks old. The weather should be cool, and it is best when done in the evening, if the sun is strong during the day. The operator grasps the bag with his left hand, and forces the testicles down to the bottom of it, so that the skin is tight over them. He then makes a slit across the bottom of the bag over each testicle, which then protrude, and the cord is cut with a knife, which should be blunt rather than otherwise.

Both ewes and lambs should be allowed to remain as quiet and undisturbed as possible for some days, and it rarely happens that any death occurs. When such does happen it is not usually confined to a few, and so far as we have seen has arisen from unnecessary disturbance by dogs, or from cold drenching rains following immediately after the operation.

Old tups may also be castrated by means of clams, as in preceding sections.

Pigs.—If the pig is not more than six weeks old, an incision is made at the bottom of the scrotum, the testicle pushed out, and the cord cut without any precautionary means whatever. But when the animal is older, there is reason to fear that bleeding to a serious extent may follow ; consequently, it will be advisable to pass a waxed string round the cord, a little above the place where the division is intended to take place. Pigs which are fat, especially aged boars, should be prepared by bleeding, cooling diet, and quiet ; and food should be withheld for at least twelve hours previous to the operation.

Spaying Heifers.—In certain parts yearling heifers are spayed to facilitate their fattening. Gamgee says, " The animal is held standing, and an assistant causes it to arch its back, thus rendering the skin over the flank tense. An incision from above, downwards, and near the rim of the pelvis, is made into the left flank, and the hand introduced downwards, and backwards, beneath the rectum, where the uterus and ovaries are found. The latter are seized, one after the other, and being brought as nearly as possible to the external opening, without dragging much on the broad ligaments, they are divided off with a spring forceps. The wound is brought together and fixed, by twisted sutures or metallic ligatures, and the animals recover readily. Low diet and careful nursing are requisite in stall-fed animals, but in countries like Australia the animals are spayed in large numbers, and allowed a free range, without endangering the success of the operation.

Spaying dairy cows has been recommended, as it causes a continuation of the secretion of milk for several years, just as it existed at the time of the operation. The best period for the operation is about six weeks after calving. Dairy cows are sometimes spayed by the vagina, but it is obviously only a professional man who can conduct the operation in a proper manner, and it is only such who should be employed to perform it.

Spaying Sows.—In spaying the sow, the animal is laid upon its left side, and firmly held by two assistants. An incision is then made into the flank, the forefinger of the right hand introduced into it, and gently turned about until it encounters and hooks hold of the right ovary, which it draws through the opening; a ligature is then passed round this one, and the left ovary felt for in like manner. The operator then severs off these two ovaries, either by cutting or tearing, and returns the womb and its appurtenances to their proper position. This being done, he closes up the womb with two or three stitches, rubs a little oil over, and the operation is at an end. Keep the animals quiet for a few days, and let the diet be cool. The best age for spaying a sow is six weeks; but some breeders take two or three litters from their sows before they operate upon them. There is, however, more risk where the operation is thus delayed.

Caponing Cock Chickens.—The chicken is placed with its left side downwards, and secured in that position by a strap which confines the wings, and by a lever which keep the legs a little asunder. An incision is then made in the side of the chicken, and held open by a pair of blunt hooks to allow the testes to be seen. These are then removed, one after the other, by means of a scoop, constructed for the purpose, which divides the membrane that covers them, and it is provided with a noose of horsehair, the action of which, operating as a saw, cuts asunder the ligatures which bind them to the back-bone. Chickens intended for capons should be operated upon when between two and three months old. Previous to the operation the chickens must be kept without food or water for about thirty-two hours, so that the bowels may be completely emptied beforehand.

FIRING.

Firing is a powerful and rapid agent employed for the purpose of producing inflammation artificially; and from its energetic action it often has the desired effect, after all other remedies have failed. In applying the cautery, casting is generally the first thing to be done. The part which is to be operated upon should previously have been shaven, or the hair clipped as short as possible. The operation consists in drawing parallel lines, about half an inch asunder, on the affected part, or a neighbouring part, as in the case of sprains of the ligaments and tendons. The instrument used for the purpose is a red hot iron with a small, smooth, rounded edge. Professor Dick cautions his readers that no part is in a fit state to be fired when the skin is hot or inflamed, and the skin should never be deeply penetrated by the iron. According to the heat of the point, so should be the velocity and lightness of touch, and a brown or yellow marking in the skin from the singeing is all that is required. After the firing the horse must be laid up for three or four days, to prevent his injuring the part. If the irritation produced is less than was intended, it may be promoted by means of blistering ointment. When it is wished to moderate or heal it, dress the parts with oil or lard.

Colonel Fitzwygram very appositely remarks that "of all notions connected with firing the most absurd is the idea that it may be useful as a preventive against future lameness. There are people," he says, "who fire, or, as they term it, 'just touch with the irons,' the inside of all their young horses' legs to prevent their having spavins, and also the posterior part of the hocks, to prevent their having curbs. What benefit they expect to derive from thus artificially exciting irritation and inflammation in a sound hock it is hard to say. Others with scarcely more reason fire the hock after curb or spavin has completely formed, although no lameness is caused. There is no sense in this. Curb, though it generally produces lameness during the process of its formation, rarely does so after the parts have consolidated. Again, if a spavin, when completely formed and consolidated, is so placed that it does not interfere with the movements of the joint, the probability is that it will never cause lameness."

SETONS AND ROWELS.

A *Seton* consists of a piece of tape or soft cord passed under a portion of the skin by the seton needle; the ends may be tied together, or the seton may be secured at its upper end by a light piece of wood, three quarters of an inch long, and at the lower end by a knot on the tape. The tape or cord must be removed, pulled up and down, every day, and sometimes twice a day, being previously lubricated with oil of turpentine or blistering plaister. The openings should also be frequently washed with lukewarm water to keep them clean, and the tape should be renewed every ten days to prevent it getting rotten.

A *Rowel* is only a seton under another form. In applying it an incision is made in the skin to the extent of about an inch, by pinching it up and cutting it with a bistoury or rowel scissors. The cellular membrane round the wound is separated to the extent of about an inch, so as to admit a pledget of tow, smeared with digestive ointment, such as oil of turpentine, one ounce; olive oil, four ounces. This soon produces a discharge, which has a tendency to relieve any deep-seated neighbouring morbid action.

/ BLISTERING. \

For this purpose an ointment is always used, of which rather more than half is well rubbed into the part to be blistered, while the remainder is thinly and equally spread over the part that has been rubbed. To ensure the full action of a blister, the hair should be removed as much as possible ; and if the legs are the parts to be acted upon, the influence of the vesicant will be more energetic and much quickened by the immersion of them for 15 or 20 minutes in warm water. To other parts fomentations may be applied, or a poultice, by which the vessels of the skin will be relaxed and rendered more susceptible of the stimulating influence of the cantharides. When there is any danger of the ointment running, and acting upon places that should not be blistered, they must be covered with a stiff ointment made of hog's lard and bee's wax, or kept wet with a little water. The bedding must be removed when the leg is blistered, and to prevent the horse slipping, the stones may be covered with a little short litter, spent tan, or saw-dust. The horse's head must be secured in such a way that he cannot reach the blister with his teeth. Put him into a warm stall, and tie his head firmly to the rack. In about six hours after the application of the blistering compound, which must be effected with friction applied the contrary way to the hair, vesication will have taken place ; and on the following day it is advisable to cleanse the part with repeated affusions of warm water, and afterwards apply the liniment of lead, or some emol- lient, such as sweet oil, or lard, by means of a soft painter's brush. Sometimes the blister becomes itchy after it is dry, and the horse rubs it ; in that case he must be tied up again ; but if he gets very tired and threatens to go down on his haunches, put a cradle on his neck, let go the head, and give him a good bed. The loose scab which forms after a blister should not be removed until it naturally peels off ; and time must be given, and no work, otherwise the horse may be blemished by the process.

Blistering is seldom resorted to in the cure of cattle, owing to the thickness of the skin; but cases sometimes occur when blisters may be of service. The following are recipes for the preparation of blisters:—

No. 1—Powdered cantharides, 2 drachms; oil of turpentine, 2 drachms; powdered euphorbium, 1 drachm; oil of origanum, 1 drachm; prepared lard, 2 ounces; mix alternately. This is a very active blister.

No. 2—Tartar emetic, 1 drachm; cantharides ointment, 2 ounces; make into an ointment. A powerful blister for horse or ox.

No. 3—Powdered croton seeds, ½ ounce; powdered cantharides, 1 ounce; oil of turpentine, 1 pint; olive oil, 1 pint; mix. A blister for the ox.

The action of a blister may be kept up for a long period by dressing it with savine ointment.

CLYSTERS, GLYSTERS, ENEMAS, OR INJECTIONS.

The common form of these agents for the horse is a liquid; occasionally, however, gaseous enemas are resorted to. The objects for which they are administered are—

1. In order to empty the bowels: thus they act as an aperient. Also to induce a cathartic to commence its action, when, from want of exercise or due preparation, it is tardy in producing the desired effect. They operate in a twofold way—first, by softening the contents of the intestines; and, secondly, by exciting irritation in one portion of the canal, which is communicated throughout the whole; hence they become valuable when the nature and progress of the disease require a quick evacuation of the bowels.

P

The usual enema is warm water, the quantity thrown up being from half a gallon to a gallon. This may be rendered more stimulating by the addition of a little common salt, or oil, or solution of aloes.

The quantity of the fluid injected should be attended to; for if this be too great, in addition to the action of the agent given, we shall have a distended state of the intestine, and a more rapid expulsion of the clyster will take place than is desirable; and, on the other hand, it should not be too small, for then the desired object will not be obtained. Various means are adopted for exhibiting enemas; the best is certainly the pump invented by the late Mr. Read; they may also be administered by means of the old ox bladder and wooden pipe; the only art required being to avoid frightening the animal, to anoint the pipe well, and to insinuate it gently before the fluid is forced up.

2. For the purpose of killing worms, which are found nidulating in the rectum and large intestines. In this case they are usually of an oleaginous nature.

3. For restraining diarrhœa: sedatives and astringents being then employed.

4. For nourishing the body, when food cannot be received by the mouth. Gruel is generally the aliment thus given.

5. For allaying spasms in the stomach and bowels. In this instance they become one of the means by which medicines are taken into the system.

The only gaseous enema is that of tobacco, which is occasionally employed in cases of severe colic, obstinate constipation of the bowels, and strangulated intestines.

FOMENTATION.

The simplest and best fomentation consists of a piece of flannel dipped in hot water, quickly wrung out, and immediately applied to the inflamed part; and to prevent the escape of heat, another piece of dry flannel should be lightly thrown over it; nor should colour be altogether disregarded.

The value of fomentations is appreciated by most practitioners; but in order to obtain the greatest amount of good from them, they should not be occasionally, but continually, employed; that is to say, the flannel should never be allowed to get cold, for then reaction is set up in the vessels of the part. When judiciously applied, they relax the capillaries, causing them to pour out a portion of their contents; thus they relieve the tension of the integument, and act almost as a local depletent, whilst the relief afforded is often communicated to deeper-seated parts. Their use is indicated during the formation of abscesses or of tumours, and after contusions and sprains. Many practitioners, finding their orders seldom attended to—namely, to keep the flannel always moist and warm—have recourse to poultices in preference; but even here a watchful eye is requisite, for if the poultice is allowed to get dry, much irritation will be created by it. The various herbs that were once recommended are now, by common consent, discarded, as tending rather to check than to augment the benefit derivable from fomentations. In neuralgic affections, however, the addition of opium or belladonna may prove of service.

THE POULTICE.

Professor Morton says that as a therapeutic agent this is usually placed among emollients. The common relaxant poultice of the Royal Veterinary College consists of bran moistened with warm water, to which is added a little linseed meal, in order to give it consistency. Care should be taken that it is kept moist; which may be effected by pouring over it, from time to time, warm water. Professor Morton is not inclined to assent to the statement that it is of little consequence whether you apply a poultice hot or cold, simply because it will soon become the same temperature as the part to which it is applied. In the effects produced by medicinal agents the impression first made by them is often of great moment. When a cold substance is applied to an inflamed surface, there will necessarily be a withdrawal of heat from it; the constringing effects of cold will then be experienced, and sometimes much pain. On the other hand, when warmth, accompanied with moisture, is resorted to, the particles are driven farther from each other, the part becomes relaxed, the distended vessels are enabled to relieve themselves, and ease is given. Sometimes, however, a cold poultice is desirable, and even ice may then be used. The effects produced by poultices likewise depend very much upon the materials composing them. Thus, as a means of softening horn in inflammation of the feet, vinegar may be used instead of water. When an astringent is desirable, a solution of alum may be added. When a disinfectant, the chloride of lime is invaluable; or a poultice containing yeast or charcoal may be employed. A boiled carrot-poultice has been found of service in ill-conditioned ulcers and irritable sores; and we have an excellent stimulating compound in the mustard cataplasm, or *sinapism* made by mixing together equal parts of mustard and linseed meal, with a sufficient quantity of boiling water so as to form a poultice. Vinegar used to be employed, but it does not increase the effects of the mustard. When speedy action is required, the flour of mustard may be used alone, made into a paste with diluted water of ammonia, and some practitioners add the oil of turpentine, which, however, is not admissible in nephritic diseases. If we are desirous of allaying irritation, opium, belladonna, or the diacetate of lead may be added to the common poultice; so that in this, as in all other branches of the practice of veterinary medicine, judgment on the part of the medical attendant is required.

ADMINISTRATION OF MEDICINES.

Horses.—Physic, in the case of horses, is frequently very much abused. Some physic their horses at stated intervals, whether medicine is required or not; others as fancy dictates, in order, as they say, to prevent the horse from getting out of order. Unless there is an absolute necessity for giving medicine, its use should be rigidly avoided; and even when the animal does appear a little out of order there are many cases when the mere preparation for physicing will produce all the effects required. "If proper and timely notice is taken of the premonitory symptoms of ailments," says Colonel Fitzwygram, "little active treatment will ever be necessary. Bran mashes instead of corn for a day or two, deprivation of hay, a cooler stable, and above all a loose box, with plenty of pure, fresh air, will probably do all that is needed, and will do it much better and more safely than physic."

But taking it for granted that the horse really requires medicine, a course of preparation is necessary before the physic is administered. This consists in giving him only cold bran mashes for a period of not less than thirty-six and up to forty-eight hours, and, if the case is slight, these mashes, being in themselves laxative, will frequently be all that is required. He may have gentle exercise morning and evening.

Physic is usually administered to the horse in the form of a ball. This is given on an empty stomach, early in the morning, followed by a drink of tepid water, or water from which the chill has been removed, by keeping it near the kitchen fire for some time; then a warm bran mash, and half an hour to an hour's exercise. Let him have as much tepid water as he will drink, and gentle exercise may be given in the afternoon, and also next day.

The physic usually begins to operate the morning after it has been given, although it may sometimes take thirty hours. When it begins to operate the horse must stand in the stable until the action ceases, or until the physic " sets,' as it is termed, and he should not be worked for at least three days after it has " set."

Considerable nicety is required in giving a ball, which should be rather of a soft consistency, and about the size of a pullet's egg. In administering it, the operator stands before the horse, which is unbound, and turned with its head out of the stall, and having a halter on it. An assistant stands on the left side, to steady the horse's head, and keep it from rising too high ; sometimes he holds the mouth, and grooms generally need such aid. The operator seizes the horse's tongue in his left hand, draws it a little out, and to one side, and places his little finger fast upon the under jaw ; with his right hand he carries the ball smartly along the roof of the mouth, and leaves it at the root of the tongue ; the mouth is closed, and the head is held till the ball is seen descending the gullet on the left side. When loath to swallow, a little water may be offered, which will carry the ball before it. Instruments have been constructed for delivering balls, but a troublesome horse should be sent at once to a veterinary surgeon. Purgatives should never be given to a horse in a weak state, as their action tends to rapidly reduce the strength still more.

An *aloetic purge* for the horse consists of from three to nine drachms of Barbadoes aloes made into a ball, with linseed meal and treacle.

Medicine is sometimes administered to the horse in the form of a drench, which operates more speedily than a ball, especially in the case of aloes in solution.

An *aloetic drench* is made by dissolving four or five drachms of aloes in a pint of hot water, with two drachms of powdered ginger and an ounce of aromatic spirits of ammonia, or half the above quantity of aloes may be dissolved, and half a pint of linseed oil or castor oil added, with two drachms of ginger. The drench should be well shaken before being given.

Drenches should invariably be administered by means of a horn, and never from a bottle, as the use of the latter frequently leads to accidents. Great care must be exercised in giving a drench: no unnecessary force should be used; and if the slightest irritation is occasioned in the windpipe, the animal's head should be let go, so that he may free himself by coughing.

Electuaries are soft preparations, in which the medicines are either tasteless or agreeable to the taste, and not too bulky. Medicine in this form is administered by means of a wooden or metal spoon, and are smeared on the tongue or the inside of the cheeks. The following is an example of an *electuary*, which may be given in sore throat in the horse:—Powdered opium, two drachms; powdered liquorice, two ounces; honey, eight ounces; mix.

Cattle.—The horn should always be used in administering medicine to cattle, unless in some cases, when it may be needful to give it by means of a hollow probang, or the fluid may be forced into the stomach by a syringe.

In administering a drench by the horn, which is the usual way, " the chief points," says Professor Gamgee, " to attend to are not to irritate the animal; always to attempt the operation from the right side; to seize hold of the upper jaw by passing the left hand over the head, and bend the latter far round to the right; the operator should stand well with his back against the animal's shoulder, propping himself up with the right leg: to do this, the animal should, especially if awkward, be against a wall on its left side;" to which we add, that the medicine should be poured gently down the throat, so that it may trickle, as it were, into the stomach.

Sheep.—These are also drenched by means of a horn, as above.

Swine.—In some cases medicine may be administered to pigs in their food or drink ; but when they refuse to take either, it must be given in the form of a drench, and " to drench a pig," Professor Gamgee remarks, " considerable care and a peculiar method must be adopted. One way consists in introducing a tolerably stout noose over the upper jaw, which is held firmly in the operator's right hand ; the pig is held between the legs ; and an assistant may aid in securing him, whilst the mixture to be given is poured out of a bottle [or horn] ; so that it trickles down the cheek, and is swallowed. If the fluid be poured in rapidly, as the pig is certain to scream, there is great danger that the fluid will pass into the windpipe, and suffocate the animal. Not unfrequently has a person, in giving medicine to a pig, observed it either suddenly or imperceptibly losing foot-hold, and dropping dead at his feet. A practice has been found to succeed admirably, which has led me to have an instrument constructed on the principle of the ' medical spoon.' The practice consists in taking an old shoe, cutting off the toe part of the upper leather, allowing the pig to suck the toe part of the sole whilst the fluid to be administered is poured into the shoe. In this way the pig absolutely sucks the mixture ; and there need be no apprehension of untoward consequences. The instrument [made in the form of a slipper, with a hollow tube handle attached] is constructed of tin. The body is covered with leather, and a tongue shaped portion of varnished leather, which the pig is allowed to take into his mouth, is made to project from the tin anteriorly. The medicine is introduced into the apparatus by the lower aperture before we commence the process of administration, and by holding the thumb over the tube at the upper part, the flow of the liquid from the instrument is much checked. The length of the instrument over its convexity, measuring from extreme points, is 14 inches ; the body is nine inches in circumference at its broadest part ; the tube is four inches in length, and two-thirds of an inch in diameter."

Mr. R. H. Dyer, V.S., Tipperary, writing on this subject in the *Irish Farmers' Gazette,* says :—" Place a small rope round the upper jaw, and raise the nose of the animal slightly upwards, just out of the level only, by passing the rope over a gate, wall, or bank ; the assistant should have the dose either in a tin, bottle, or drenching horn, which may be used for the purpose, the dose being gently allowed to flow into the mouth and swallowed. If the pig should be taken between the legs of an assistant, and the head elevated, as is usually done, ten chances to one suffocation will take place."

Dogs.—See page 187.

INOCULATION FOR PLEURO-PNEUMONIA.

Inoculation has been useful in many instances as a preservative against pleuro-pneumonia. Gamgee describes the operation as follows :—" A portion of diseased lung is chosen, and a bistoury or needle made to pierce it, so as to become charged with the material consolidating the lung, and this is afterwards plunged into any part, but more particularly towards the point of the tail. If operated severely, and higher up, great exudation occurs, which spreads upwards, invades the areolar tissue round the rectum and other pelvic organs, and death soon puts an end to the animal's excruciating suffering. If the operation be properly performed *with lymph that is not putrid, and the incisions are not made too deep,* the results of the operation are limited to local exudation and swelling, general symptoms of fever, and gradual recovery. The most common occurrence is sloughing of the tails."

It is important to notice the words put in Italics, as we have known serious losses to follow when putrid matter was used, and the incisions made deeper than what was necessary.

THE PULSE.

Professor Dick states that the natural pulse of the horse is from 35 to 45 beats in the minute; others from 32 to 40. It is usually quicker in young horses than aged animals. The pulse of the ox or cow is also from 35 to 45; that of the ass, 48 to 54; sheep, 70 to 79; dog, 90 to 100.

APPENDIX.

CATTLE.

The following was accidently omitted in its proper place :—

THE WHITES (LEUCORRHŒA).

Symptoms.—This is a discharge of mucus or white slimy matter, and sometimes a yellow or green coloured matter, from the womb and vagina, accompanied with loss of milk and loss of appetite, weakness, and a general unthrifty appearance. Should cows be put to the bull when so affected, abortion will probably ensue. If the malady is unchecked, diarrhœa will make its appearance, and render it more difficult to treat the disease. It is not contagious.

Treatment.—Give internally the following dose :—Common turpentine, $\frac{1}{2}$ ounce ; hydrochloric acid, 1 scruple ; powdered ginger, 2 drachms ; decoction of oak bark, 10 ounces. Mix. Give also, as an injection, alum dissolved in water. Good food and pure water, with open air exercise, is useful ; and give also tonics, such as gentian and ginger, half an ounce of each, powdered, in warm beer. Oilcake, bruised oats, or oatmeal gruel should enter largely into the dietary.

In addition to the particulars given under the different heads in the body of this work, the following remarks on certain substances employed in preserving or restoring the health of animals, will be found useful. We are indebted for them to Professor Morton's " Veterinary Pharmacy."

CARBOLIC ACID.

As an ordinary disinfectant for stables, cow-sheds, &c., one gallon of carbolic acid may be mixed with ten gallons of white-wash. For the floors, a much more diluted solution, one part to a hundred, will be found effectual. In cancerous sores and ill-conditioned wounds, canker and thrush in the foot of the horse, and foot-rot in sheep, a mixture of one part of carbolic acid and four to six parts of glycerine may be employed. It is useful also in sore backs and shoulders, broken knees, &c. Apply with a brush.

OINTMENT OF CARBOLIC ACID.

Carbolic acid, second quality, eight ounces; lard, three pounds. The addition of a little white wax will be necessary in hot weather. A valuable application for sloughing and unhealthy wounds. Mixed with two parts of sulphur, the ointment is of great benefit in mange in the horse and dog, and likewise for scab in sheep; and it also allays the itchings in many of those affections of the skin which do not depend upon parasites.

BLACK HELLEBORE.

For cattle, black hellebore has been for a long time employed; it being inserted as a seton in the dewlap. Some practitioners make use of the leaves, but the fresh root is by far the most active. It quickly produces much swelling, which is followed by suppuration.

In some parts of Germany it is used in a similar way for dogs labouring under distemper. Vomition is caused by it, and relief is thus afforded. The following formula for an active digestive ointment for cattle has been communicated by Mr. E. J. Sparrow, who speaks highly of it :—

Take of leaves of black hellebore and hogs' lard, of each equal parts. Boil together till the leaves become crisp, then strain off, and add common turpentine equal in weight to the ointment obtained. Mix.

OINTMENT OF THE BINIODIDE OF MERCURY.

Take of biniodide of mercury ... 1 part.
Lard ... ... 8 parts.

Intimately mix.

This, applied to sores, is a stimulant and detergent, and as such it may be employed when they have taken an unhealthy action. But its more general use is as a counter-irritant to the skin, which it powerfully excites, sometimes inducing intense erysipelatous inflammation, accompanied with much pain, and followed by desquamation of the cuticle. By it absorption is also very much facilitated; and hence it has been found of service in splents, curbs, incipient spavins, enlarged bursæ, thickening of the integument, indurated tumours, and abnormal growths. In some cases, from the susceptibility of the skin to be irritated by it, the quantity of the biniodide requires to be lessened to one half. The application of the ointment should be accompanied with friction; and when soreness has been induced by it, and a vesicular eruption appears, its use should be abstained from for a time; but as soon as these effects have passed off, it may be again and again resorted to. It may also be applied to farcy ulcers.

Some practitioners prefer the ointment of the biniodide of mercury to that of cantharides as a blistering compound, considering it to be more efficacious; while, instead of removing the hair, if judiciously applied, they say it promotes its growth.

COMMON SALT.

Salt is an invaluable tonic and alterative. It may be given in doses of one or two ounces, sprinkled over the animal's provender, which will induce him to eat it with avidity. It gently stimulates the stomach and alimentary tube, thus increasing the power of the digestive organs, by which the tonicity of the system is restored. In large doses it is said to be an anthelmintic and cathartic.

When the horse, the cow, or the sheep is becoming convalescent, and the natural and sanitary stimulus of wholesome food will produce a more certain as well as a safer impulse to the discharge of the natural functions than any medicine can afford, a little salt, or salt and water, sprinkled on the food will be an admirable excitant. On all the ruminantia the influence of this agent is marked; nor are the carnivora less benefited by it. In fact, it appears to be the natural stimulus to the digestive organs of all animals.

Mr. Youatt states that there is no medicine for the rot in sheep which is of the slightest avail in which culinary salt is not the principal ingredient. Also, as a purgative, it is second only to Epsom salts in the first instance; and, whether from the effect of the change of the medicine, or of some chemical composition or decomposition which takes place, it is the surest aperient that can be given when the sulphate of magnesia has failed. It may be administered in the same doses as that agent. Being a tonic as well as a purgative, it is on this account, perhaps, somewhat objectionable in the early stages of fever. It is a vermifuge which in cattle seldom fails. To the dog it has been exhibited for the same purpose, being administered in repeated doses till the stomach no longer rejects it. The pig would seem to be highly susceptible of its influence, since many instances of this animal having been poisoned by it are recorded in the archives of veterinary medicine. It is frequently used as an adjunct to clysters. Externally applied, dissolved in water, in the proportion of one pound to a gallon, it is employed as a stimulant for chronic sprains. For although while undergoing solution the temperature of the water is lowered, yet rarely is it the case that only during this time is it employed; but the solution is kept in the stable until it acquires the same temperature as the air of the stable, and then being applied it becomes a stimulant. If made use of only while the heat of the water is being abstracted, so as to cause the salt to pass from the state of solid to that of liquid, it would, of course, be a refrigerant.

SULPHUR.

As a therapeutic agent, it is extolled as a laxative and alterative. As the former, it is rarely used for the horse, although it has been said to possess anthelmintic properties. The French veterinarians state that a pound of it act as a poison on the horse, destroying life. Mr. Morton has given this quantity more than once, and it was followed by much intestinal irritation, and a relaxed state of the fæces only; soon the kidneys were called into increased action by it, and the pulse became accelerated. He has been informed that a large quantity of pulverised roll sulphur having been given to two horses, one of them died from it, and on a *post-mortem* examination, the intestines were found incoated with the agent and violently inflamed. The other horse was purged by it for some days.

Mr. R. H. Dyer states that a continental writer has advocated the employment of the fumes of sulphur for chest affections in the human subject, and that he has so used it with some success. He adds—"The administration of sulphur to the horse in ordinary coughs and colds, I have great faith in. In many instances I have prescribed the remedies for cough, and they have signally failed; and then I have found marked beneficial effects by administering ounce doses of this drug in a bran mash for a few nights. If common sulphur was worth a shilling an ounce, it would meet with more favour as a medicine."

Sulphur and oil of tar is a very effectual remedy for many skin diseases, both in horses and dogs.   The only secret in its use is, to cover the entire animal with it, instead of employing it upon *parts* of the skin only.

When the bowels of cattle and sheep have been excited to action by the sulphate of magnesia, this action is, to a moderate extent, and with perfect safety, kept up by subsequent doses of sulphur ; the quantity for the former being from six to eight ounces, and for the latter, from two to three ounces.   It may also be advantageously joined with other purgatives.   As an alterative it is usually administered in combination with the nitrate of potassa and the sulphide of antimony.

Externally applied, it is valuable in many cutaneous affections, and deserving of general employment.   For instance, in mange in the horse it may be added to the compound liniment of tar ; and for cattle, sheep, and the dog, it is the basis of all the compounds used for this annoying disease in them.

INDEX.

Q